Mathematics in Mind

SpringerBriefs in Cognitive Mathematics

Series Editor

Stacy A. Costa, ON Institute for Studies in Education,
University of Toronto, Toronto, ON, Canada

Marcel Danesi, Department of Anthropology, University of Toronto,
Toronto, ON, Canada

SpringerBriefs present concise summaries of cutting-edge research and practical applications across a wide spectrum of fields. Featuring compact volumes of 50 to 125 pages, the series covers a range of content from professional to academic. Briefs are characterized by fast, global electronic dissemination, standard publishing contracts, standardized manuscript preparation and formatting guidelines, and expedited production schedules.

Typical topics might include:

- A timely report of state-of-the art techniques
- A bridge between new research results, as published in journal articles, and a contextual literature review
- A snapshot of a hot or emerging topic
- An in-depth case study
- A presentation of core concepts that students must understand in order to make independent contributions

SpringerBriefs in Cognitive Mathematics showcases research in the study of math cognition as well as the relation of math to other faculties, such as language, music, and art, among others. As a subseries of **Mathematics in Mind**, the research published in this series falls under the aegis of the Fields Cognitive Science Network, which brings together mathematicians and cognitive scientists to explore the nature of mathematics from various interdisciplinary angles. All works are peer-reviewed to the highest standards of scientific literature.

Marcel Danesi

Image Schema Theory and Mathematical Cognition

Springer

Marcel Danesi
Department of Anthropology
University of Toronto
Toronto, ON, Canada

ISSN 2522-5405　　　　　　　ISSN 2522-5413　(electronic)
Mathematics in Mind
SpringerBriefs in Cognitive Mathematics
ISBN 978-3-031-85413-2　　　ISBN 978-3-031-85414-9　(eBook)
https://doi.org/10.1007/978-3-031-85414-9

© The Editor(s) (if applicable) and The Author(s), under exclusive license to Springer Nature Switzerland AG 2025

This work is subject to copyright. All rights are solely and exclusively licensed by the Publisher, whether the whole or part of the material is concerned, specifically the rights of translation, reprinting, reuse of illustrations, recitation, broadcasting, reproduction on microfilms or in any other physical way, and transmission or information storage and retrieval, electronic adaptation, computer software, or by similar or dissimilar methodology now known or hereafter developed.
The use of general descriptive names, registered names, trademarks, service marks, etc. in this publication does not imply, even in the absence of a specific statement, that such names are exempt from the relevant protective laws and regulations and therefore free for general use.
The publisher, the authors and the editors are safe to assume that the advice and information in this book are believed to be true and accurate at the date of publication. Neither the publisher nor the authors or the editors give a warranty, expressed or implied, with respect to the material contained herein or for any errors or omissions that may have been made. The publisher remains neutral with regard to jurisdictional claims in published maps and institutional affiliations.

This Springer imprint is published by the registered company Springer Nature Switzerland AG
The registered company address is: Gewerbestrasse 11, 6330 Cham, Switzerland

If disposing of this product, please recycle the paper.

Preface

What parts of the brain are activated when we add or subtract numbers? Which parts are involved in allowing us to understand and utilize abstractions such as equations and matrices? Are the same neural circuits and processes involved in such concepts the same ones involved in language comprehension and use? Why are diagrams so helpful, if not indispensable, in grasping and representing mathematical ideas? As Aristotle famously wrote, all such questions about human mentation can be combined into one primary conundrum: How do we know what we know?

This monograph will address Aristotle's conundrum with regard to mathematics, taking a specific perspective to do so, based on George Lakoff and Rafael Núñez's book, *Where Mathematics Comes From: How the Embodied Mind Brings Mathematics Into Being* (2000), in which the two researchers presented a coherent, albeit still controversial, view of how mathematicians come to invent, discover, and use their ideas by deploying the same cognitive mechanisms that underlie abstract thought in language and other faculties—metaphor, metonymy, and analogy. The ideational roots of such thinking, as the two researchers maintained, are *image schemas*, defined as mind forms that crystallize from the everyday experiences of things, such as how we move along paths, how we put and organize objects in containers of different sizes, how we balance things in order to reach a state of equilibrium, and so on. These experiences are abstracted by the brain into structural forms, which, when they become embedded in neural circuitry, constitute sources of unique ideas such as number lines, coordinate systems, sets, and infinity, among many others. Ever since Lakoff and Núñez's book, image schema theory has become a target of theoretical interest, research, and debate among psychologists, mathematicians, educators, and others who are engaged in the study of *mathematical cognition*, a term that came into wide use near the end of the 1990s in reference to how we learn, invent, and use mathematics.

A fundamental premise within the image schema theoretical framework is that math cognition is an offshoot of the same neural-conceptual system that undergirds language and other faculties, such as drawing. Research subsequent to Lakoff and Núñez's book has been providing substantive support for this very premise. My goal is to elaborate and expand upon image schema theory, as a model of math cognition that resonates intuitively with math educators and mathematicians alike. So, for instance, the intuitive physical sense derived from looking *up* and *down* is extracted by the brain to form an image schema of verticality that guides our understanding of an array of abstract concepts in mathematics (and other systems) in terms of how we represent them: for example, a curve plotted on a graph that turns in an upward direction is conceived as portraying "increase" of some kind, while a curve that turns downwards as representing "decrease" instead. Now, the question becomes: How can we know that an image schema such as the verticality one is real in any psychological sense? The supportive evidence comes from three sources—research on the learning of mathematics, representations of mathematical ideas such as diagrams, which reflect image-schematic structure, and AI research incorporating them as modeling devices—all of which will be discussed in the final chapter. There is now even a formal language to describe image-schematic structure, called appropriately, Image Schema Language, which is able to represent different layers of image schemas—the result of a project headed by Maria Hedblom, a leading researcher in computer science at the University of Jönköping in Sweden.

I have designed this book for all kinds of readers, not only for those in the math cognition field. It is based in large part on my own research into image schemas in mathematics education and on various theoretical studies that I have conducted on metaphorical reasoning in mathematics over the years. As James Geary so keenly put it in a 2011 post to his blog, *The Macmillan Dictionary blog* (macmillandictionaryblog.com/metaphors-in-mind): "Metaphors hide in plain sight, and their influence is largely unconscious. We should mind our metaphors, though, because metaphors make up our minds." Hopefully, I will be able to show, or at least argue cogently, that image schemas, as the sources of conceptual metaphors in mathematics, "make up the mind" as it learns and does mathematics, not just tangentially, but intrinsically.

Toronto, On, Canada Marcel Danesi
2024

Contents

1 **The Starting Point: Lakoff and Núñez** 1
   1.1   Introduction 1
   1.2   Studying Mathematical Cognition 2
   1.3   The Number Line 5
   1.4   The Basic Metaphor of Infinity 8
   1.5   Grounding and Linking Metaphors 11
   1.6   Mathematics and Language 13
   1.7   Plausibility 16
   References 18

2 **Image Schema Theory** 23
   2.1   Introduction 23
   2.2   Image Schemas 25
   2.3   A Typology 32
   2.4   The Gestalt Background 35
   2.5   Stasis–Motion–Force 38
   2.6   System Thinking 41
   2.7   The Etymology of Schema 44
   References 45

3 **Related Processes** 49
   3.1   Introduction 49
   3.2   Mapping 50
   3.3   Framing 53
   3.4   Layering 57
   3.5   Blending 60
   3.6   Clustering and Radial Networks 65
   3.7   Looping 67
   References 69

4	**Learning, Diagrams, and AI**	73
	4.1 Introduction	73
	4.2 Learning	74
	4.3 Diagrams	80
	4.4 AI Modeling	85
	4.5 Validation	88
	References	90
Index		95

Chapter 1
The Starting Point: Lakoff and Núñez

1.1 Introduction

Studying mathematical cognition with diverse disciplinary tools, including psychology, neuroscience, AI, linguistics, and anthropology, among others, started in earnest at the turn of the millennium (Butterworth, 1999; Dehaene, 1997, 2004; Devlin, 2000, 2005; Gallistel & Gelman, 2005; Lakoff & Núñez, 2000). Several trends within this field have since emerged, making it a varied and fascinating domain of investigation that has concrete implications for the study of mind more generally. This introductory chapter will focus on a particular line of inquiry within the field—the relation of mathematics to metaphorical cognition, which is traced to George Lakoff and Rafael Núñez's 2000 book, *Where Mathematics Comes From: How the Embodied Mind Brings Mathematics Into Being*, especially since image schema theory—the particular topic of interest of this book—was first incorporated by the authors in that book to explain math cognition—a theory that however was initially developed and fine-tuned in the 1980s for the study of language (for example, Johnson, 1987; Lakoff, 1987; Lakoff & Johnson, 1980, 1999).

The point of view taken here is that, since the Lakoff and Núñez book, it has become virtually unavoidable to take into some consideration the role of metaphorical reasoning in shaping math cognition—a role that connects mathematics to language as integrated, rather than separate, faculties of mind. The Lakoff–Núñez book caused a stir when it first came out, receiving both positive and negative reactions to it (for example, Davis, 2005; Henderson, 2002), but what is of specific relevance here is that it has produced a pattern of testable insights into what happens in the mind as we do mathematics, thus suggesting that image schema theory is a viable construct for gaining access into that mind. Of course, there are different ways to approach math cognition, and these will be referenced in specific ways throughout this book. The differences of viewpoints in the field bring out the many challenges involved in studying math cognition meaningfully (see Klein et al., 2023). But

Lakoff and Núñez's book is still significant because it was the "starting point" for the use of image schema theory in the study of math cognition—hence its particular importance to the present discussion.

A guiding question that will come up throughout this book, implicitly and explicitly, is whether there is such a thing as "thinking purely in numbers," which, if true, would imply a form of cognition in the case of mathematics separate from linguistic cognition, or whether such thinking is part of a broader form of neural processing that involves language and other faculties, such as drawing. In other words: Can there be mathematics without language? There is actually relevant research that indicates that there might indeed be a separate mathematics system in the brain. In one study, Varley et al. (2005) found that three individuals with brain damage retained remarkable mathematical skills but were unable to use language to communicate. Even the understanding of a mathematical phrase such as "seven minus two" was beyond their comprehension. Findings such as this one have been interpreted by some in the field as indicating that mathematics and language are processed in different parts of the brain and thus as constituting separate cognitive faculties. A putative and initial response to this finding is that subjects with brain damage may not be ideal ones for resolving the mathematics-versus-language debate, especially since, as will be argued throughout this book, it appears that both are embedded on the same image-schematic foundation of thought. Studies involving brain-damaged subjects invariably utilize verbal prompts or instructions to get the subjects to perform mathematical activities—a fact that in itself indirectly suggests that the two faculties of language and mathematics are intertwined in some way.

1.2 Studying Mathematical Cognition

As a first illustrative example of what image schema theory might entail, consider the outline visual features of a physical path, such as a sidewalk on a city street. First, we notice that it has a linear structure. Then, it is obvious that when we start walking on it, we do so in terms of a starting point, from where we move forward to a desired end point or goal via a series of intermediate points. We can, of course, also turn around and walk the other way if need be. Now, these features are encoded by the experiential centers of the brain, gradually becoming extrapolations in the form of a recurring mental "portrait" of path structure, which incorporates thought forms such as movement from place to place, a starting point, a goal for engaging in the movement, and a series of intermediate points on the way to reaching the end-goal, on which we may stop to take stock of where we are. This type of brain-based portraiture congeals into an image schema. Acquired in childhood, image schemas are formed on the basis of observable characteristics of the environment. They are the brain's version of a perceptual summary or paraphrase of some physical or sensory experience that recurs in everyday life.

Once an image schema is formed and embedded in the neural substrate, it becomes an ontological guide for imagining and expressing abstract ideas across human

1.2 Studying Mathematical Cognition

faculties. Consider the path schema as it manifests itself in the conceptualization of such abstractions as "life" and "memory." In language, for instance, its unconscious presence can be seen to guide metaphorical utterances such as "There is a lot of life ahead of you" (a path forward) and "I am looking back on my life" (a path backward). Now, this very same image schema can be seen in the ideation of the number line, first imagined by John Wallis in 1685, as a construct which portrays numbers as points on a linear path moving forward (increasing) and backward (decreasing). Rather than digits, Wallis used letters to indicate what he described as "advancing and retreating" from the starting point A (now represented as zero on the line) (Fig. 1.1).

It is remarkable to note that a specific mathematical idea and verbal metaphors related to topics such as "life" and "memory" are based on and guided by the same image schema of the path. Now, Lakoff and Núñez's main claim is that image schemas such as the path one underlie mathematical and verbal thinking in the same way. Integers, for instance, are subtended ontologically by a container image schema, that is, as objects in containers—a schema that also undergirds the notion of sets, with numbers perceived as different kinds of objects assembled in the same types of containers (integers, primes, rationals, imaginary, etc.). The same schema manifests itself in verbal metaphors such as "I am full of memories," "There are not enough hours left to do this properly," and so on. While these might appear disconnected conceptually on the surface, below it they are formed on the basis of the same container schema.

The critical question becomes: Is the origin of the number line truly traceable to the path image schema or set theory to the container image schema? Or is this just an imaginative, but psychologically naïve way, to describe the origin of such mathematical constructs? Hedblom et al. (2015), among others, have answered this question convincingly in the positive, pointing out that the relevant psychological literature on childhood development does indeed support the notion of image schema, even if indirectly. The path image schema, for example, has been found to be "one of the first image schemas to be acquired in early infancy as children are immediately exposed to movement from a range of objects" (Hedblom et al., 2015: 26). It is thus becoming an increasingly tenable position, based on relevant research, to say that there is a common image-schematic basis on which language and mathematics are built (Danesi, 2016), given that the same schemas seem to mark the acquisition of concepts in both.

Before Lakoff and Núñez, a major researcher in the field, Stanislas Dehaene (1997), asserted that "number sense," the biologically determined ability to grasp and use basic numerical concepts (Dantzig, 1930), is located in the inferior parietal

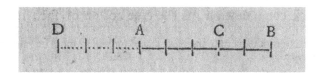

Fig. 1.1 Wallis's number line (1685)

cortex, which suggests that it is organized in a different region than "word sense," as it can be called for the sake of argument. But the separate-faculty argument was addressed and contradicted by the many illustrations used in *Where Mathematics Comes From* which support the integration, not separation, of number sense and word sense. The aim of this opening chapter is to take a retrospective look at the plausibility of Lakoff and Núñez's approach to math cognition—a view that is still not shared by some influential researchers in the field. One cogent argument that has been used against their approach is the documented presence of number sense in some nonhuman animals, who have no human language, or word sense, as such. Another argument comes from research which seems to indicate that young children can count, using finger gestures, before they can speak (Gallistel & Gelman, 2005; Gilmore et al., 2018; Gilmore, 2023a for an overview). The meaning of such differences in research outcomes and theories vis-à-vis the Lakoff–Núñez model will be discussed throughout this book.

A way of explaining, and perhaps reconciling, the discrepancies that are felt to exist between the Lakoff–Núñez model and other approaches is to adopt a triadic distinction based on the chronological phases characterizing the understanding and learning of mathematics, which can be called generically as "pre-math," "math," and "mathematics," as Alexander (2012) has aptly noted. In this book, reference to these three phases will be signaled by quotation marks. "Pre-math" refers to an innate and intuitive stage of math understanding. It is the phase in which a primitive number sense emerges, which does indeed show up with counting gestures, as relevant research shows. It also involves the ability to differentiate between small quantities, to understand how counting objects can be put in sequences, and to identify shapes as being bigger or smaller. "Math" refers instead to the subsequent phase during which number sense gives way to arithmetical competence—the ability to consciously understand numbers as abstract mental objects. It is during this phase that arithmetical competence is acquired and developed in tandem with linguistic competence, a fact that is corroborated by studies which show that the same image schemas manifest themselves in childhood linguistic and mathematical behaviors (Hedblom et al., 2015). Finally, "mathematics" refers to the phase when formal mathematical skills are acquired via some training intervention in a specific learning context. It also refers to the discipline itself, with its own professional culture, its research agendas, and epistemologies.

As Alexander emphasizes, the flow of understanding goes from the "pre-math" intuitive stage to the conceptually formative "math" stage and, in many instances, to the advanced "mathematics" stage. Similarly, Lakoff and Núñez refer to the "pre-math" phase as the period of "innate arithmetic," a period when notions of size are subitized, that is, immediately recognized. Subitizing is a term that was coined by Jean Piaget (1952) in reference to the child's early ability to look at a small set of objects and instantly recognize how many there are (more or less) without counting them. It is a key cognitive event in the early development of math cognition (Gelman & Gallistel, 1978).

It can be suggested here, as an aside, that the notion of image schema is akin to the idea of existential graphs, put forth by American philosopher and mathematician

1.3 The Number Line

Fig. 1.2 An existential graph

Charles Peirce (1931–1956, vol. 2: 398–433, vol. 4: 347–584), which he described as diagrams representing "moving pictures of thought." As Kiryushchenko (2012: 122) has observed, for Peirce, diagrams portray in their physical form or design how we "experience a meaning visually as a set of transitional states, where the meaning is accessible in its entirety at any given 'here and now' during its transformation." An example of an existential graph is the one that Peirce used to show the relation A > B (Roberts, 2009; Stjernfelt, 2007) (Fig. 1.2).

The diagram represents the relation A > B by putting A above B, as the "greater" one, and B below as the "smaller" one below. In terms of Lakoff and Núñez's approach, this can be seen to be a diagrammatic version of the verticality schema mentioned briefly above, which implies the conceptual metaphor, *higher is greater* and *lower is smaller*. As Louis Kauffman (2001: 80) has observed: "Peirce's Existential Graphs are an economical way to write first order logic in diagrams on a plane." Diagrams are pictures of thought transferred to some surface, such as paper or a screen, reflecting the image-schematic structure and features of thought (Chap. 4). They are thus visual evidence of what arguably goes on in the mind.

1.3 The Number Line

Starting in the late 1970s, a theoretical framework emerged in psychology and linguistics which viewed abstract cognition as being built on a process called "conceptual metaphorizing" (Honeck & Hoffman, 1980; Lakoff & Johnson, 1980; Ortony, 1979)—a framework adopted and adapted by Lakoff and Núñez to explain math cognition, laying the groundwork for subsequent research on image schema theory in the math cognition field (Berch et al., 2018; Bockarova et al., 2012; Danesi, 2019; Geary et al., 2015; Schlimm, 2013). In this approach, a single metaphor, such as "She is a courageous lion," is not seen as an isolated rhetorical figure of speech, but rather as a result of an underlying thought formula, called a conceptual metaphor, *humans are animals*, which is an unconscious form of abstract understanding based on metaphorical cognition. In the formula, *humans* is called the target domain and *animals* the source domain. "She is a courageous lion" is just one token manifestation of the conceptual metaphor at the level of discourse, among many others: "My friend is an eagle," I am a social butterfly," "Beware of human foxes," and so on.

Consider the number line. Wallis (1685: 265) wrote that he devised it as a concrete diagram to show that addition and subtraction were conceptually akin to someone walking forward and backward along a linear path. Wallis thus prefigured, with his explanation, the view of the number line as a conceptual metaphor, resulting from the image schema of a path which, itself, is derived from the physical experience of locomotion and the unconscious absorption of the features of physical paths and of the motion of walking on them, as discussed (Khatin-Zadeh et al., 2023). To cite Núñez (2017), the number line is "a conceptual tool that allows for numbers to be conceived as locations along a line mapping numerical difference onto difference in spatial extension." In microcosm, this episode in mathematical history provides philological (text-based) corroboration of the validity of the notion of conceptual metaphorizing as guided by image-schematic thought. Now, the result of introducing the number line into mathematics has hardly been decorative or just illustrative, as Wallis seems to have believed. It has led to the view of number as a discrete entity (a point on the line), increasing or decreasing in value according to its location on an unending linear configuration—a view (literally) that forms the basis of concepts such as infinity, negative numbers, zero, and others (as will be discussed). As Talmy (1996: 217) has aptly observed, the number line suggests "fictive motion," whereby there is "a continuous linear entity emerging from the front of some object and moving steadily away from it."

However, it cannot be taken for granted that this construct will be understandable to everyone—a possibility that is often used against the notion of image schema and the Lakoff–Núñez approach more generally. This is so because it emanates initially as an artifact in the cognitive domain of "mathematics" and thus may need concrete explanatory interventions or representational devices employed illustratively at the levels of "pre-math" and "math." Young children, for instance, appear to have difficulties with this artifact, because it presents numbers as abstractions (Cross et al., 2009). For this reason, devices such as bead strings have been used as transitional pedagogical tools to get children to conceptualize the number line construct more abstractly, because these are physically "graspable" as representations of numbers laid out in a connected sequential pattern (the string) (Peyser & Bobo, 2022). Children at the "pre-math" stage, and at an early "math" stage as well, need appropriate instruction that will allow them to grasp abstractions such as the number line. A relevant study by Link et al. (2013) compared the pedagogical outcomes of using two types of number lines on first graders. One involved conventional training, asking the children to indicate the position of a number on a tablet-based number line. The other, which they called an "embodied condition," involved instructing the children to walk along a physical line on the floor until they reached the target number's position. With the latter method, the researchers found that skill in single-digit addition improved significantly for the walking group as compared with the tablet group.

Educational studies are crucial for examining the plausibility of image schema theory, as will be discussed further in Chap. 4. Mathematical education is, in fact, one of the main areas to which image schema theory has been applied most fruitfully (Danesi, 2007; Presmeg, 2005; Yee, 2017). The viability of an "image-schema-based pedagogy" is evidenced in elementary school math classrooms every day, even if the teacher grasps the notion in an intuitive way. Manipulatives such as beads are used commonly to impart concepts of quantity and numeration. By putting the beads into containers of varying sizes (larger versus smaller), and a numerical name applied to each container, corresponding to the quantity of objects in it, the conceptual understanding of numeration appears to emerge spontaneously in children.

1.3 The Number Line

Another potential problem with the number line concept is that it was conceived in a culture where geometrical linearity is found in various forms and structures—rulers, thermometers, scales, and so on. Disentangling culture from mathematical invention is challenging and vigorously debated as a major factor is shaping the learning of mathematics (Dehaene, 1997; Dehaene et al., 2008; Núñez, 2017; Pitt et al., 2021; Siegler, 2016). Number lines appear to be incomprehensible to those who do not come from the same cultural rearing context where linearity is a design element in artifacts (Edmonds-Wathen, 2012). Nonetheless, by simply introducing the number line into a specific learning context, regardless of the background culture of the learners, its basic image-schematic principles seem to be highly understandable. It all depends, as alluded to above, on the type of instructional method used, which can raise the path schema to consciousness. Lakoff and Núñez (2000: 279) were aware of the role of culture in metaphorical conceptualization. However, as they also noted (Lakoff & Núñez, 2000: 356), "once mathematical ideas are established in a worldwide mathematical community, their consequences are the same for everyone, regardless of culture." Moreover, there is anthropological evidence that the number line concept exists as an intuitive, rather than explicit, mathematical one across cultures, as Núñez et al. (2012) point out: "It has been proposed that the number line is based on a spontaneous universal human intuition, rooted directly in brain evolution, that maps number magnitude to linear space with a metric. To date, no culture lacking this intuition has been documented."

Interestingly, a study by Alcaraz-Carrión et al. (2022) found that the number line concept comes out visibly in the accompanying gestures people make when speaking about addition and subtraction. So, when discussing addition the subjects of their study would move a hand to the right from the body, mirroring an imaginary right-ward movement of numbers on the number line, whereas when the same subjects were discussing subtraction, they would use a hand movement oriented to the left of the body. The researchers describe their findings as follows:

> Most gestures about addition and subtraction were produced along the lateral or sagittal axes. When people spoke about addition, they tended to produce lateral, rightwards movements or movements away from the body. When people spoke about subtraction, they tended to produce lateral, leftwards movements or movements towards the body. This co-speech gesture data provides evidence that people activate two different metaphors for arithmetic in spontaneous behavior: arithmetic is motion along a path and arithmetic is collecting objects.

This type of finding is consistent with the general findings on co-speech communicative behavior (vocal language and accompanying gestures). Research by Adam Kendon (2004) and David McNeill (1992, 2005), among others, has established a structural–conceptual relation between gestures and words during speech, with the former mirroring the semantic-conceptual content of the latter in gestural behavior. This line of inquiry suggests that gestures are integrated, not just illustrative, components of vocal communication—exhibiting images that cannot be shown overtly in speech, as well as images of what the speaker is thinking about.

Now, once a metaphorical construct becomes part of the discipline of "mathematics," it comes to constitute the source of further metaphorical conceptualizations—a phenomenon called layering in this book (Chap. 3). So, the conceptual move from one-dimensional (linear) space to two-dimensional (and, eventually, higher-dimensional) spaces is suggested by an image-schematic mapping resulting in two number lines crossing at right angles, producing the coordinate system. Without the initial number line metaphor, however, this would hardly have crystallized as a possibility. It was sparked arguably by the image-schematic structure of crossing paths. The coordinate system is, in fact, a conceptual blend of two number lines which, as David et al. (2017) point out, shows how location and magnitude converge into a single system (Fauconnier & Turner, 2002). In this framework, the number line, after being conceived, is seen as constituting a new source domain that is mapped onto the target domain of the plane, consisting of intersecting lines at right angles, which are "two distinct cognitive structures with fixed correspondences between them" (Lakoff & Núñez, 2000: 48).

1.4 The Basic Metaphor of Infinity

The mathematical notion of infinity as numbers moving along an unending linear path is a classic example of layering, that is, of how one conceptual metaphor, the number line, suggests other ontologically related concepts via subsequent metaphorical mappings. Lakoff and Núñez explained infinity in terms of a so-called *Basic Metaphor of Infinity* (BMI), reflecting the notion of "adding one more" to any linear sequence of numbers. This same notion is the ontological basis of proof by induction, which states that if some condition holds for the $(n + 1)$th case, given n, then it holds infinitely, because we can add the $(n + 2)$nd case, the $(n + 3)$rd case, and so on—one case at a time, until we decide to stop, which Lakoff and Núñez called "completion."

The BMI can be seen to be behind key discoveries such as Galileo's (1638) demonstration that there are as many square integers as there are positive integers, even though the squares are themselves a subset of the set of integers. Galileo showed this by simply laying out the two sets on separate number lines and then matching the corresponding points on the two lines ad infinitum (Fig. 1.3).

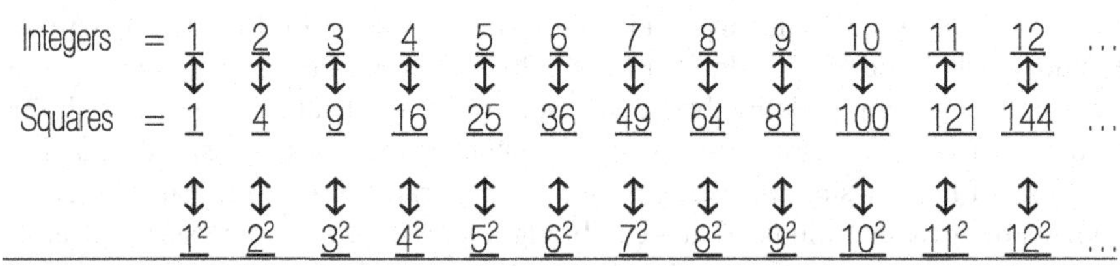

Fig. 1.3 Galileo's paradoxical demonstration

1.4 The Basic Metaphor of Infinity

This shows that no matter how far we go along the two lines there will never be a gap, and thus that the numbers on one line correspond to the numbers on the other line—in more technical terms, they have the same cardinality. In 1874, Cantor showed that the same one-to-one correspondence method could be used to prove that the same cardinality holds between the integers and numbers raised to any power (Fig. 1.4).

Cantor then went on to develop a whole new branch of mathematics based on the correspondence method of proof (Tiles, 2004). The point here is that the insight for this new form of mathematical reasoning and proof came from using the number line conceptual metaphor in a new way, leading to the discovery that there were different orders of infinity—a truly mind-boggling result (Chap. 2). The Galilean–Cantorian type of demonstration and the novel mathematics that it entailed make sense because they tapped into the image-schematic structure of the number line extended by the BMI in a specific conceptual form.

The BMI construct has been critiqued on several counts (Wagner, 2013), but in counter-position to these critiques, it can be seen to provide, at the very least, a theoretical framework for explicating notions such as cardinality and the kinds of demonstrations that were devised by Galileo and Cantor. As Godino et al. (2011: 250) emphasize, a portrayal such as the BMI of a "limitless collection of objects" may be a universal one, translating countability into a conceptual system:

> As we have freedom to invent symbols and objects as a means to express the cardinality of sets, that is to say, to respond to the question, how many are there?, the collection of possible numeral systems is unlimited. In principle, any limitless collection of objects, whatever its nature may be, could be used as a numeral system.

Now, as mentioned, once a metaphorical thought system is established, it becomes the source of further mappings, resulting in an expansion of mathematics. In a lecture he gave at the Field's Institute in 2011, Lakoff explained how Kurt Gödel's famous incompleteness proofs (1931) were guided by Cantor's diagonal proof method, constituting a particular instantiation of the unconscious operation of

Integers	=	1	2	3	4	5	6	7	8	9	10	11	12	...
		$\updownarrow$	$\updownarrow$	$\updownarrow$	$\updownarrow$	$\updownarrow$	$\updownarrow$	$\updownarrow$	$\updownarrow$	$\updownarrow$	$\updownarrow$	$\updownarrow$	$\updownarrow$	
Powers	=	1^3	2^3	3^3	4^3	5^3	6^3	7^3	8^3	9^3	10^3	11^3	12^3	...
		$\updownarrow$	$\updownarrow$	$\updownarrow$	$\updownarrow$	$\updownarrow$	$\updownarrow$	$\updownarrow$	$\updownarrow$	$\updownarrow$	$\updownarrow$	$\updownarrow$	$\updownarrow$	
		1^4	2^4	3^4	4^4	5^4	6^4	7^4	8^4	9^4	10^4	11^4	12^4	...
		$\updownarrow$	$\updownarrow$	$\updownarrow$	$\updownarrow$	$\updownarrow$	$\updownarrow$	$\updownarrow$	$\updownarrow$	$\updownarrow$	$\updownarrow$	$\updownarrow$	$\updownarrow$	
		1^5	2^5	3^5	4^5	5^5	6^5	7^5	8^5	9^5	10^5	11^5	12^5	...
		...	...	...	...	...	...	...	...	...	...	...	...	
		$\updownarrow$	$\updownarrow$	$\updownarrow$	$\updownarrow$	$\updownarrow$	$\updownarrow$	$\updownarrow$	$\updownarrow$	$\updownarrow$	$\updownarrow$	$\updownarrow$	$\updownarrow$	
		1^n	2^n	3^n	4^n	5^n	6^n	7^n	8^n	9^n	10^n	11^n	12^n	...

Fig. 1.4 Cantor's demonstration

the BMI (see Danesi, 2011)—that is, the same method of one-to-one demonstration based on number line reasoning (diagonal in this case) (Bou et al., 2015). The diagonal method, Lakoff asserted, must have been in Gödel's mind when he showed that within a diagonal layout of symbols or propositions, there is one that does not fit in. While this is a liberal reduction of Lakoff's argument, the point is that the BMI, as Lakoff claimed, was the likely mental image that influenced Gödel to devise his famous proofs of incompleteness—the first proof showed that in a system of propositions there will be at least one proposition within it that is "true" but "unprovable," the second demonstrated that the system cannot indicate its own consistency. The first theorem was based on a modified version of the Liar Paradox (Chap. 3), replacing "this sentence is false" with "this sentence is not provable," called the "Gödel sentence G." To put it in conceptual metaphor terms, the sentence G cannot be mapped anywhere in the collection of symbols within a layout.

As Rafael Núñez (2005: 1717) explained in an article he wrote several years after the publication of *Where Mathematics Comes From*, responding to critiques, the BMI can perhaps be renamed more appropriately the Basic Mapping (rather than Metaphor) of Infinity, since this describes concretely the ontological substance of Cantor's and Gödel's proofs:

> [Cantor's] analysis is based on the Basic Metaphor of Infinity (BMI). The BMI is a human everyday conceptual mechanism, originally outside of mathematics, hypothesized to be responsible for the creation of all kinds of mathematical actual infinities, from points at infinity in projective geometry to infinite sets, to infinitesimal numbers, to least upper bounds. Under this view "BMI" becomes the Basic Mapping of Infinity.

The BMI is an example of what Lakoff and Núñez called a "linking metaphor," discussed in the next section, a metaphorical mechanism that is behind concepts such as points at infinity, limits, infinite intersections, etc. It allows us, in other words, to conceptualize "potential infinity" as we go along an endless path and stop at points selected as resultant states. As Lakoff and Núñez (2000: 160) put it:

> What results from the BMI is a metaphorical creation that does not occur literally: a process that goes on and on indefinitely and yet has a unique final resultant state, a state "at infinity." This metaphor allows us to conceptualize "potential" infinity, which has neither end nor result, in terms of a familiar kind of process that has a unique result. Via the BMI, infinity is converted from an open-ended process to a specific, unique entity.

Since the BMI is a linking metaphor, it is thus a layered one—a metaphor of a previous metaphor, the number line in this case. It can itself become a source domain that is mapped further onto different kinds of possibilities, including that of "approaching" something, such as zero, which entails the spatial view of distances on a linear path becoming smaller and smaller until they reach a completion, termed limit in mathematics (Lakoff & Núñez, 2000: 155). In order to have a limit, therefore, one must "reach" it at the infinite step (at least potentially). While it may be important sometimes to know what happens at the limit (as when considering continuity), this is not necessary when finding the limit itself.

Despite critical reviews of the BMI, such as those by Auslander (2001) and Gold (2001), which point out infelicities in some of the mathematics used by Lakoff and

Núñez, the BMI remains intuitively resonant to this day, describing how the diagrammatic proofs and demonstrations of Galileo, Cantor, and Gödel were remarkably consistent in form and conceptualization because of this metaphor. We would be hard-pressed to find another viable psychological explanation for their proofs. The mathematical details of the BMI can be adjusted, of course, but overall it does not crumble under the critiques, at least in my view.

1.5 Grounding and Linking Metaphors

Lakoff and Núñez (2000) posit two main types of conceptual metaphors underlying mathematical concepts, which they call *grounding* and *linking* metaphors. The former are the result of primitive or first-order image schemas that go into the ontological constitution of source domains that are then mapped onto initial mathematical concepts directly—a first-order image schema is one that is derived from actual lived experiences, such as walking along a path or putting things into containers. An example of a conceptual metaphor that is guided by this order of image-schematic thought is the *arithmetic is object collection* conceptual metaphor, based on the container image schema, which envisions numbers as "objects in imaginary collections," derived from the experience of "putting objects together, using measuring sticks, and moving through space" (Lakoff & Núñez, 2000: 102). Linking metaphors, on the other hand, operate at a more abstract level, mapping previous metaphors onto new domains, as discussed above with regard to Gödel's demonstrations. The BMI is such a metaphor, allowing us to conceptualize infinite sets as "special cases of a single general conceptual metaphor in which processes that go on indefinitely are conceptualized as having an end and an ultimate result" (Lakoff & Núñez, 2000: 158).

Lakoff and Núñez identified four grounding metaphors of arithmetic, which can be paraphrased here for illustrative purposes:

1. *Object collection*: This refers to how we conceive of arithmetical operations and patterns in terms of the container image schema. For example, $3 + 2 = 5$ involves combining a collection of three objects and a collection of two objects into a new collection of five objects. This will happen every time—a process that Lakoff has called invariance or inference-preservation (1979). Subtraction involves reverse image-schematic thinking, whereby objects are taken from a collection and the leftover is counted.
2. *Object construction*: This reflects the image schema of numbers as constructed objects rather than collections, built up of parts, as for example, fractions. This schema intuitively guides the use of pedagogical explanations and demonstrations in elementary school mathematics such as the following: If you cut a pie in half, you will get two equal pieces. Each of these is "one of two." Now, if you cut the same pie into three equal pieces, each one is "one of three" and so on. The addition of the separate pieces adds up to the whole pie.

3. *Measuring stick*: Measuring involves visualizing numbers as one-dimensional objects on a line laid out in some sequence, reflecting the path image schema that undergirds the number line concept.
4. *Motion along a path*: This is the source of constructs such as positive and negative numbers, the latter being objects on the number line moving "the other way" from zero (leftward rather than rightward). A linear path might even go on to the horizon, and beyond, which undergirds the idea of mathematical infinity.

There are many more examples that could be used to illustrate the metaphors above, but these need not concern use here. Suffice it to say that these four grounding metaphors cover virtually all of arithmetic from its elementary school uses to its sophisticated number theoretical forms. As Moronuki (2018) observes with regard to the cognitive effects of these metaphors:

> The combination of these four metaphors give us a lot to work with—numbers as multidimensional tactile objects, numbers as spatial relationships, and so on. However, doing arithmetic physically with one of these source frames behaves the same as arithmetic in any of the others…[from which], we formalize and make precise the laws of arithmetic.

The path and container image schemas seem to be behind the generation of a large number of mathematical concepts, including graph and set theory (to be discussed subsequently). To reiterate, the former elicits a mental picture of numbers as points along a linear path and the latter as three-dimensional objects that are contained in, or can be put in, collections—mental pictures that are easily expressed in the form of diagrams such as number lines and circles as sets (Chap. 4). These are qualitatively different, and yet, as Lakoff and Núñez point out, we can identify and formalize the ways in which they behave analogously. So, if we can lay out a horizontal number line, then, by crossing it with a vertical one at right angles, called a zero point, we can then conceptualize numbers as occurring as ordered pairs in the two-dimensional space created by the intersection of the lines (the coordinate plane), or in the case of complex numbers, as "real" and "imaginary" number lines crossing in the same way to produce the complex number system (Chap. 4). Even non-discrete numerical objects (irrational numbers and π, for example) can be located on the same line via construction techniques, transforming the line into a model of the real numbers, R (Fig. 1.5).

These schemas thus form the first-order level of cognitive understanding. They are everywhere in mathematics. As Lakoff and Núñez asserted in their book, it was the blending of linearity and containment that must have even inspired George Boole (1854) to come up with the idea that binary arithmetic could express all states

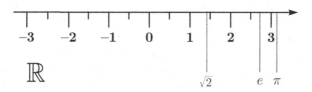

Fig. 1.5 The real number line

1.6 Mathematics and Language

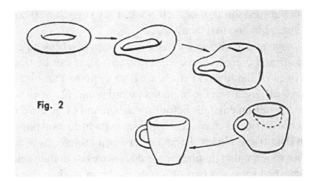

Fig. 1.6 The topology of a donut and coffee mug (Barr, 1964)

in a system of logic. Binary numbers can be laid out in the same linear way as other numbers and can be used to represent presence or absence (1 or 0) in a system—that is, as present or absent from a container. This paved the way for Claude Shannon (1948) to provide a way of understanding the logical operations of computer systems in terms of "on-off" circuits.

Image-schematic structure can even be seen as being cognitively operative in topological mathematics, in which figures can be stretched into other figures and still retain their essential structure. In this case, the relevant schema is the force one, to be discussed in the next chapter. This implies that geometric forms can be "forced" into new ones, so to speak, via some transformational process. For example, a circle is topologically equivalent to a stretched ellipse, and a donut is topologically equivalent to a coffee cup. By stretching and pinching, but not tearing or joining anything, the donut can be transformed into a coffee mug, with its "hole" corresponding to the handle of the coffee mug. A drawing of the transformation from donut to coffee mug is the following classic one by Stephen Barr in his book, *Experiments in Topology* (1964) (Fig. 1.6).

The donut and mug share the property of invariance (Lakoff, 1979), mentioned above, which makes them topologically equivalent. Other processes which display invariance in some way involve specific types of mappings, discussed further in Chap. 4 (Hofstadter, 1979; Hofstadter & Sander, 2013).

1.6 Mathematics and Language

The fact that the same image schemas undergirding linguistic concepts also underlie mathematical ones is indirect evidence, but nonetheless strong evidence, that the two are intertwined cognitively—recall the path schema manifesting itself in verbal utterances such as "Life is a long journey" and in the form and layout of the number line. Such correlations between language and mathematics are not coincidental, as

Lakoff and Núñez argued throughout their book. They suggest a common ground of understanding shared by the two faculties.

Linking language and mathematics has always been of a bone of contention cast against the Lakoff and Núñez approach. As discussed, it can be traced mainly to work in three lines of inquiry: (1) studies which indicate that elemental number sense is organized differently in the brain than word sense; (2) work which suggests that in childhood counting emerges before speaking; and (3) research which shows that number sense exists in nonhuman species which do not possess language. Already before Lakoff and Núñez, Brian Butterworth brought forth evidence in his 1999 book, *What Counts*, that number sense does not exist in the same brain region where word sense resides—it is part of a separate brain module, located in the left parietal lobe, which also controls the movement of fingers, explaining why we count instinctively with them, without being taught to do so. In the field of childhood development, studies such as the one by Libertus et al. (2009) are suggestive of the chronologically separate emergence of number and word senses. The researchers presented 7-month-old infants with familiar and novel number concepts as electroencephalogram measures of their brain activity were recorded. The result indicated that the brains of infants can detect numerical novelty independently of words for the new forms, seemingly providing evidence for an innate and separate number sense in infancy. Such results have been corroborated by other developmental studies. However, in this same domain of studies, this very pattern of results is differentiated according to "early" and "mature" number sense phases (Libertus & Brannon, 2009), which is highly relevant to the discussion at hand, since the early phase coincides with what has been called the "pre-math" stage here and the mature one with the "math" stage—a differentiation that is critical in discussing number sense versus arithmetical competence (discussed further in Chap. 4).

Like Butterworth, in *The Number Sense* (1997), Stanislas Dehaene put forth a cogent argument for differentiating between the two senses (number and word), basing it on experimental evidence which indicates that animals such as rats, pigeons, raccoons, and chimpanzees can perform the same kind of simple calculations that human infants can. When a rat is trained to press a bar to receive a food reward, the number of presses approximates a Gaussian distribution with a peak around 8 or 16 bar presses. Lions in the African Savannah are able to distinguish between 1 and 5 lion calls even if these artificially emitted in their habitats by researchers (McComb et al., 1994). If a single lioness hears three calls from an unknown lion source, she will leave, but if she is with four of her sisters, they will go out together and explore the source of the calls. Not only can lions tell when they are "outnumbered," but also that they will act on the basis of numerically significant signals. Since animals do not possess language, it can be deduced that number sense, as evidenced by its manifestation in nonverbal animals, does not require language to sustain it.

Various case studies of brain-damaged patients have also come forth to support the separation of mathematics from language (Ardila & Rosselli, 2002; Butterworth et al., 2011; Dehaene et al., 2003; Isaacs et al., 2001). Patients with acalculia (the inability to do simple arithmetic operations), who might read 14 as 4, have difficulty associating number concepts with the words used to encode them. In effect, they appear to have difficulty matching the language to the math. Dyscalculia patients, who have an inability to carry out simple calculations, associated with impairments in the intraparietal sulcus, can continue to speak but cannot do simple mathematics.

One of the first psychologists to study the development of number sense in children was French psychologist Jean Piaget (1926, 1952, 1969; Piaget & Inhelder, 1969). In a set of classic experiments, Piaget wanted to determine if the abstract concept of conservation varied according to the age of the child—he used the term *conservation* in reference to the ability to recognize that a specific quantity will remain the same despite adjustments to the contents of a container or to its shape. In one of the experiments (Piaget & Szeminska, 1941), Piaget presented two glass containers, Glass 1 and Glass 2, filled with liquid to the same height, to a very young child—a child who had reached what today would be identified as the "mature" or "math" phase (above), a post-neonate stage that dovetailed with the child's grasp of elementary language. He then asked the child if the two containers had the same amount of liquid in them, to which the child would automatically answer "yes." Piaget then poured the liquid from Glass 2 into another glass, Glass 3, which was lower and wider in shape but had the same volume capacity of both Glass 1 and 2, asking the child if the amount of liquid was still the same, to which the child responded, this time, that the amount was not the same, saying that Glass 3 contained more liquid. Now, when the same experiment was conducted with an older child, the subject concluded that the amount of liquid was in fact still the same in Glass 3. As Skemp (1971: 154) pointed out with regard to Piaget's work, while number sense may start out as an instinctive ability at the preverbal stage, as the child becomes verbal, the two senses—word and number—develop in tandem, as the words for numbers become indistinguishable from the number concepts: "counting is so much a part of the world around them that children learn to recite number-names not long after they learn to talk."

In two relevant books, *The Math Gene* (2000) and *The Math Instinct* (2005), Keith Devlin also suggests that there must be an innate number sense; otherwise, no one would be able to count instinctively, no matter what language they speak or what culture they were reared in. He raises an important question: Why can we speak easily, but not do "math" so easily (in most cases)? The answer is that people can and do "math" but do not recognize that they are doing it, even if with difficulty. As Devlin argues, the brains of our prehistoric ancestors must have had both number and word sense as they gained consciousness and eventually created early cultures. But those brains could hardly have imagined how to multiply 15 by 36 or to prove Fermat's Last Theorem. It was language that came forth to help the human species reflect upon such complex ideas, because it allowed humans to frame them conceptually. This suggests an evolutionary partnership between the language and mathematics.

This is the gist of the implicit question that Lakoff and Núñez asked initially: How does the linkage between number and word sense manifest itself? Following their arguments, the manifestation is evidenced in the shared image-schematic mechanisms that typify verbal and mathematical concepts, as discussed several times, and as will be elaborated in the remainder of this book. So, while there is evidence that number and word sense may be located in different brain regions in infancy, the two can hardly be separated after that phase, as both the evolutionary and developmental research literatures suggest.

1.7 Plausibility

Another major critique leveled at Lakoff and Núñez is that they frame their theory as all-encompassing one, rather than presenting it as one plausible way, among others, to explain math cognition. As Voorhees (2004: 85) pointed out in his review of *Where Mathematics Comes From*, "a frustrating aspect of the book is the authors' underlying hostility to views of mathematics other than their own." However, other approaches, as I read them, are also presented as all-inclusive, not just by Lakoff and Núñez. As Camilla Gilmore, a prominent researcher on math cognition, has put it (Gilmore, 2023b): "we do not have a shared viewpoint on what mathematics is. As a result, it is difficult to bring together findings from different studies and we don't know how a set of varied skills, processes and knowledge combine to allow individuals to be mathematical." She goes on to make a set of cogent suggestions for bringing together the different theoretical strands:

- Determine if relevant how different research findings fit together.
- Identify outstanding questions for future research.
- Refine the choice of mathematical measures.
- Provide a shared technical language to discuss these issues.

If nothing else, it can be asserted that, ever since the publication of *Where Mathematics Comes Form*, the notion of image schema as a neurological source of mathematical ideas can no longer be ignored within the math cognition field. It provides a plausible explication of what goes on in the mind as we invent, do, and learn both language and mathematics. Today, there is a huge dataset of research findings, which, as Lakoff and Dodge (2005) observed, is providing strong supportive evidence for this construct as a pivotal "structure of experience, thought, and language" (for an overview of relevant research, see Rohrer, 2005).

Another critique of the Lakoff–Núñez approach is that it might simply be nothing more than an analogical argument of how metaphor facilitates the learning of mathematics, as Winter and Yoshimi (2020) have maintained: "the evidence collected in the embodied mathematics literature is inconclusive: It does not show that abstract mathematical thinking is constituted by metaphor; it may simply show that abstract thinking is facilitated by metaphor." But one could also argue that there is no theory of the mathematical mind that is not based on metaphorical-analogical reasoning. In a significant 1941 book written for the general public by Courant and Robbins, titled *What is Mathematics?* their answer to their own question was in fact analogical—that is, they illustrate what mathematics looks like and what it does by comparing it to other human activities, allowing us to come to our own conclusions as to what mathematics is with regard to these activities. So, despite some early critiques of *Where Mathematics Comes From*, its main principle that abstract

1.7 Plausibility

mathematical thought is metaphorical in origin has become a widely used one in the field of math cognition, even if their approach continues to be ignored by some key researchers in the field. Bonnie Gold (2001) encapsulated the essence of the Lakoff–Núñez approach as follows: "while we can subitize numbers as large as 4, we can't subitize 4 + 3. For this we need metaphor."

Lakoff and Núñez's approach has also revived the Platonic-versus-constructivist debate. Do we discover mathematics or do we invent it and then discover that it works? Was $\sqrt{2}$ "out there" ready to be discovered, or was it produced inadvertently by the manipulation of mathematical forms? As is well known, Plato (in the *Meno*, 380 BCE) believed that mathematical ideas pre-existed in the mind and that is why we recognize them when we come across them. Some now find this perspective difficult to accept, leaning toward constructivism, or the idea that mathematical ideas are constructed to tell us what we need to know about the world (for example, Davis & Hersh, 1986; Hersh, 1997, 1997, 2014). Constructivism is the point of view taken by Lakoff and Núñez. But, as Berlinski (2013: 13) suggests, the Platonic view is not so easily dismissible:

> If the Platonic forms are difficult to accept, they are impossible to avoid. There is no escaping them. Mathematicians often draw a distinction between concrete and abstract models of Euclidean geometry. In the abstract models of Euclidean geometry, shapes enjoy a pure Platonic existence. The concrete models are in the physical world.

Without the Platonic models, the concrete ones would bear little interest to us mathematically. Moreover, there might be a neurological basis to the Platonic view. As neuroscientist Pierre Changeux (2013: 13) muses, Plato's trinity of the Good (the aspects of reality that serve human needs), the True (what reality is), and the Beautiful (the aspects of reality that we see as meaningful) is actually consistent with the notions of modern-day neuroscience:

> So, we shall take a neurobiological approach to our discussion of the three universal questions of the natural world, as defined by Plato and by Socrates through him in his *Dialogues*. He saw the Good, the True, and the Beautiful as independent, celestial essences of Ideas, but so intertwined as to be inseparable…within the characteristic features of the human brain's neuronal organization.

This means that our brain is predisposed to look for reality, but that it might be a reality of our making, as Lakoff and Núñez ultimately suggest. Moreover, Plato's models mean that we should never find faults within our systems of mathematical knowledge, for then it would mean that the brain is faulty. As it turns out, this is what Gödel's (1931) theorems implied. But then, if mathematics is faulty, why does it lead to demonstrable discoveries, both within and outside of itself? Perhaps the connection between the brain, the body, and the world will always remain a mystery, since the brain cannot really study itself.

As Sackett (2014: 198–199) has aptly observed, our mind "works in ways that are mathematical," but we would be remiss "if we think the mathematical mind is just about mathematics." While many details related to image schema theory brought forth by Lakoff and Núñez continue to be debated, there is little doubt that the relation between number and word sense is more than an isomorphism. The claim to be made in the remainder of this book is that the link between the two emerges when image schemas manifest themselves in exact, not isomorphic, ways as we do mathematics.

References

Alcaraz-Carrión, D., Alibali, M. W., & Valenzuela, J. (2022). Adding and subtracting by hand: Metaphorical representations of arithmetic in spontaneous co-speech gestures. *Acta Psychologica, 228*, 103624. https://doi.org/10.1016/j.actpsy.2022.103624

Alexander, J. (2012). On the cognitive and semiotic structure of mathematics. In M. Bockarova, M. Danesi, & R. Núñez (Eds.), *Semiotic and cognitive science essays on the nature of mathematics* (p. 134). Lincom Europa.

Ardila, A., & Rosselli, M. (2002). Acalculia and dyscalculia. *Neuropsychology Review, 12*, 179–231.

Auslander, J. (2001). Embodied mathematics. *American Scientist, 89*. https://doi.org/10.1511/2001.28.0

Barr, S. (1964). *Experiments in topology*. Dover.

Berch, D. C., Geary, D. C., & Koepke, K. M. (Eds.). (2018). *Language and culture in mathematical cognition*. Academic.

Berlinski, D. (2013). *The king of infinite space: Euclid and his elements*. Basic Books.

Bockarova, M., Danesi, M., & Núñez, R. (Eds.). (2012). *Semiotic and cognitive science essays on the nature of mathematics*. Lincom Europa.

Boole, G. (1854). *An investigation of the laws of thought*. Dover.

Bou, F., Corneli, J., Gómez-Ramírez, D., Smaill, E., Maclean, A., & Pease, A. (2015). The role of blending in mathematical invention. In *Proceedings of the Sixth International Conference on Computational Creativity* (pp. 55–62). Association for Computational Creativity.

Butterworth, B. (1999). *What counts: How every brain is hardwired for math*. Free Press.

Butterworth, B., Varma, S., & Laurillard, D. (2011). Dyscalculia: From brain to education. *Science, 332*, 1049–1053.

Cantor, G. (1874). Über eine Eigneschaft des Inbegriffes aller reelen algebraischen Zahlen. *Journal für die Reine und Angewandte Mathematik, 77*, 258262.

Changeux, P. (2013). *The good, the true, and the beautiful: A neuronal approach*. Yale University Press.

Courant, R., & Robbins, H. (1941). *What is mathematics? An elementary approach to ideas and methods*. Oxford University Press.

Cross, C. T., Woods, T. A., & Schweingruber, H. (Eds.). (2009). *Mathematics learning in early childhood*. The National Academies Press.

Danesi, M. (2007). A conceptual metaphor framework for the teaching mathematics. *Studies in Philosophy and Education, 26*, 225–236.

Danesi, M. (2011). George Lakoff on the cognitive and neural foundations of mathematics. *Fields Notes, 11*(3), 14–20.

Danesi, M. (2016). *Language and mathematics*. Mouton de Gruyter.

Danesi, M. (Ed.). (2019). *Interdisciplinary perspectives on mathematical cognition*. Springer.

Dantzig, T. (1930). *Number: The language of science*. Macmillan.

David, E. J., Roh, K. H., & Sellers, M. E. (2017). The role of visual reasoning in evaluating complex mathematical statements. In T. Fukawa Connelly, K. Keene, & M. Zandieh (Eds.), *Proceedings of the 20th Annual Conference on Research in Undergraduate Mathematics Education, San Diego, California* (pp. 93–107).

Davis, E. (2005). Review of Where mathematics comes from. *Journal of Experimental & Theoretical Artificial Intelligence, 17*, 305–315.

Davis, P. J., & Hersh, R. (1986). *Descartes' dream: The world according to mathematics*. Houghton Mifflin.

Dehaene, S. (1997). *The number sense: How the mind creates mathematics*. Oxford University Press.

Dehaene, S. (2004). Arithmetic and the brain. *Current Opinion in Neurobiology, 14*, 218–224.

Dehaene, S., Piazza, M., Pinel, P., & Cohen, L. (2003). Three parietal circuits for number processing. *Cognitive Neuropsychology, 20*, 487–506.

References

Dehaene, S., Izard, V., Spelke, E., & Pica, P. (2008). Log or linear? Distinct intuitions of the number scale in Western and Amazonian indigene cultures. *Science, 320*, 1217–1220.

Devlin, K. J. (2000). *The math gene: How mathematical thinking evolved and why numbers are like gossip*. Basic.

Devlin, K. J. (2005). *The math instinct: Why you're a mathematical genius (along with lobsters, birds, cats and dogs)*. Thunder's Mouth Press.

Edmonds-Wathen, C. (2012). Spatial metaphors of the number line. In J. Dindyal, L. P. Cheng, & S. F. Ng (Eds.), *Mathematics education: Expanding horizons*. Mathematics Education Research Group of Australasia Inc.

Fauconnier, G., & Turner, M. (2002). *The way we think: Conceptual blending and the mind's hidden complexities*. Basic.

Galilei, G. (1638). *Dialogues concerning two new sciences* (p. 1914). Macmillan.

Gallistel, C. R., & Gelman, G. (2005). Mathematical cognition. In K. Holyoak & R. Morrison (Eds.), *The Cambridge handbook of thinking and reasoning* (pp. 559–588). Cambridge University Press.

Geary, D. C., Berch, D. C., & Koepke, K. M. (Eds.). (2015). *Evolutionary origins and early development of number processing*. Elsevier.

Gelman, R., & Gallistel, C. (1978). *The child's understanding of number*. Harvard University Press.

Gilmore, C. (2023a). Understanding the complexities of mathematical cognition: A multi-level framework. *Journal of Experimental Psychology, 76*, 1953–1972.

Gilmore, C. (2023b). *What is mathematics? How should we make sense of mathematical cognition research*. Centre for Mathematical Cognition. https://blog.lboro.ac.uk/cmc/2023/05/31/what-is-mathematics-how-should-we-make-sense-of-mathematical-cognition-research/

Gilmore, C., Göbel, S. M., & Inglis, M. (2018). *An introduction to mathematical cognition*. Routledge.

Gödel, K. (1931). Über formal unentscheidbare Sätze der Principia Mathematica und verwandter Systeme, Teil I. *Monatshefte für Mathematik und Physik, 38*, 173–189.

Godino, J. D., Font, V., Wilhelmi, R., & Lurduy, O. (2011). Why is the learning of elementary arithmetic concepts difficult? Semiotic tools for understanding the nature of mathematical objects. *Educational Studies in Mathematics, 77*, 247–265.

Gold, B. (2001). Review of *Where mathematics comes from*. Mathematical Association of America. https://old.maa.org/press/maa-reviews/where-mathematics-comes-from-how-the-embodied-mind-brings-mathematics-into-being

Hedblom, M. M., Kutz, O., & Neuhaus, F. (2015). Choosing the right path: Image schema theory as a foundation for concept invention. *Journal of Artificial General Intelligence, 6*, 21–54.

Henderson, D. W. (2002). Review of Where does mathematics come from. *Mathematical Intelligencer, 24*, 75–78.

Hersh, R. (1997). *What is mathematics really?* Oxford University Press.

Hersh, R. (2014). *Experiencing mathematics*. American Mathematical Society.

Hofstadter, D. (1979). *Gödel, Escher, Bach: An eternal golden braid*. Basic Books.

Hofstadter, D., & Sander, E. (2013). *Surfaces and essences: Analogy as the fuel and fire of thinking*. Basic.

Honeck, R. P., & Hoffman, R. R. (Eds.). (1980). *Cognition and figurative language*. Lawrence Erlbaum Associates.

Isaacs, E. B., Edmonds, C. J., Lucas, A., & Gadian, D. G. (2001). Calculation difficulties in children of very low birthweight: A neural correlate. *Brain, 124*, 1701–1707.

Johnson, M. (1987). *The body in the mind: The bodily basis of meaning, imagination and reason*. University of Chicago Press.

Kauffman, L. K. (2001). The mathematics of Charles Sanders Peirce. *Cybernetics & Human Knowing, 8*, 79–110.

Kendon, A. (2004). *Gesture: Visible action as utterance*. Cambridge University Press.

Khatin-Zadeh, O., Farsani, D., Hu, J., & Marmolejo-Ramos, F. (2023). The role of perceptual and action effector strength of graphs and bases of mathematical metaphors in the metaphorical processing of mathematical concepts. *Frontiers in Psychology, 14*, 1178095.

Kiryushchenko, V. (2012). The visual and the virtual in theory, life and scientific practice: The case of Peirce's quincuncial map projection. In M. Bockarova, M. Danesi, & R. Núñez (Eds.), *Semiotic and cognitive science essays on the nature of mathematics* (pp. 46–59). Lincom Europa.

Klein, E., Zamarian, L., & Kaufmann, L. (2023). Challenges in understanding numerical learning: Editorial for *Brain Sciences* special issue "neurocognitive signatures of math (learning) across the lifespan and their interrelation with other aspects of cognition and emotion". *Brain Sciences, 28*, 420. https://doi.org/10.3390/brainsci13030420

Lakoff, G. (1979). The contemporary theory of metaphor. In A. Ortony (Ed.), *Metaphor and thought*. Cambridge University Press.

Lakoff, G. (1987). *Women, fire and dangerous things: What categories reveal about the mind*. University of Chicago Press.

Lakoff, G., & Dodge, E. (2005). Image schemas: From linguistic analysis to neural grounding. In B. Hampe (Ed.), *From perception to meaning: Image schemas in cognitive linguistics*. Mouton de Gruyter.

Lakoff, G., & Johnson, M. (1980). *Metaphors we live by*. Chicago University Press.

Lakoff, G., & Johnson, M. (1999). *Philosophy in the flesh: The embodied mind and its challenge to western thought*. Basic.

Lakoff, G., & Núñez, R. (2000). *Where mathematics comes from: How the embodied mind brings mathematics into being*. Basic Books.

Libertus, M. E., & Brannon, E. M. (2009). Behavioral and neural basis of number sense in infancy. *Current Directions in Psychological Science, 18*, 346–351.

Libertus, M. E., Pruitt, L. B., Woldorff, M. G., & Brannon, E. M. (2009). Induced alpha-band oscillations reflect ratio-dependent number discrimination in the infant brain. *Journal of Cognitive Neuroscience, 21*, 2398–2406.

Link, T., Moeller, K., Huber, S., & Fischer, U. (2013). Walk the number line—An embodied training of numerical concepts. *Trends in Neuroscience and Education, 2*, 74–84.

McComb, K., Packer, C., & Pusey, A. (1994). Roaring and numerical assessment in contests between groups of female lions, *Panthera leo*. *Animal Behavior, 47*, 379–387.

McNeill, D. (1992). *Hand and mind: What gestures reveal about thought*. University of Chicago Press.

McNeill, D. (2005). *Gesture & thought*. University of Chicago Press.

Moronuki, J. (2018). *The unreasonable effectiveness of metaphor*. Argumatronic. https://argumatronic.com/posts/2018-09-02-effective-metaphor

Núñez, R. (2005). Creating mathematical infinities: Metaphor, blending, and the beauty of transfinite cardinals. *Journal of Pragmatics, 37*, 1717–1741.

Núñez, R. (2017). Is there an evolved capacity for number? *Trends in Cognitive Science, 17*, 409–424.

Núñez, R., Cooperrider, K., & Wassmann, J. (2012). Number concepts without number lines in an indigenous group of Papua New Guinea. *PLoS One, 7*, e35662. https://doi.org/10.1371/journal.pone.0035662

Ortony, A. (Ed.). (1979). *Metaphor and thought*. Cambridge University Press.

Peirce, C. S. (1931–1958). In C. Hartshorne, P. Weiss, & A. W. Burks (Eds.), *Collected papers of Charles Sanders Peirce* (Vol. 1–8). Harvard University Press.

Peyser, E., & Bobo, J. (2022). Linking number sense to linear space. *Mathematics Teacher: Learning and Teaching PK-12, 115*, 113–121.

Piaget, J. (1926). *The language and thought of the child*. Routledge & Kegan Paul.

Piaget, J. (1952). *The child's conception of number*. Routledge and Kegan Paul.

Piaget, J. (1969). *The child's conception of the world*. Littlefield, Adams & Co.

Piaget, J., & Inhelder, J. (1969). *The psychology of the child*. Basic Books.

Piaget, J., & Szeminska, A. (1941). *La genèse du nombre chez l'enfant*. Delachaux: Niestle.

References

Pitt, B., Ferrigno, S., Cantlon, J. F., Casasanto, D., Gibson, E., & Piantadosi, S. T. (2021). Spatial concepts of number, size, and time in an indigenous culture. *Science Advances, 7*, eabg4141. https://doi.org/10.1126/sciadv.abg4141

Plato (380 BCE). *Meno*. Internet Archive. classics.mit.edu/Plato/meno.html

Presmeg, N. C. (2005). Metaphor and metonymy in processes of semiosis in mathematics education. In J. Lenhard & F. Seeger (Eds.), *Activity and sign* (pp. 105–116). Springer.

Roberts, D. D. (2009). *The existential graphs of Charles S. Peirce*. Mouton.

Rohrer, T. (2005). Image schemata in the brain. In B. Hampe (Ed.), *Image schemas in cognitive linguistics* (pp. 165–196). Mouton de Gruyter.

Sackett, G. (2014). The lines that make the clouds: The essence of the mathematical mind in the first six years of life. *North American Montessori Teachers Association Journal, 39*, 197–215.

Schlimm, D. (2013). Conceptual metaphors and mathematical practice: On cognitive studies of historical developments in mathematics. *Topics in Cognitive Science, 5*, 283–298.

Shannon, C. E. (1948). A mathematical theory of communication. *Bell Systems Technical Journal, 27*, 379–423.

Siegler, R. S. (2016). Magnitude knowledge: The common core of numerical development. *Developmental Science, 19*, 341–361.

Skemp, R. R. (1971). *The psychology of learning mathematics*. Penguin.

Stjernfelt, F. (2007). *Diagrammatology: An investigation on the borderlines of phenomenology, ontology, and semiotics*. Springer.

Talmy, L. (1996). Fictive motion in language and "ception". In P. Bloom, M. Peterson, L. Nadel, & M. Garrett (Eds.), *Language and space* (pp. 211–276). MIT Press.

Tiles, M. (2004). *The philosophy of set theory: An historical introduction to Cantor's paradise*. Dover.

Varley, R. A., Klessinger, N. J., Romanowski, C. A., & Siegal, M. (2005). Agrammatic but numerate. *Proceedings of the National Academy of Sciences, 102*, 3519–3524.

Voorhees, B. (2004). Embodied mathematics: Comments on Lakoff & Núñez. *Journal of Consciousness Studies, 11*, 83–88.

Wagner, R. (2013). A historically and philosophically informed approach to mathematical metaphors. *International Studies in the Philosophy of Science, 27*, 109–135.

Wallis, J. (1685). *Treatise of algebra*. Royal Society.

Winter, B., & Yoshimi, J. (2020). Metaphor and the philosophical implications of embodied mathematics. *Frontiers in Psychology, 11*, 569487. https://doi.org/10.3389/fpsyg.2020.569487

Yee, S. P. (2017). Students' and teachers' conceptual metaphors for mathematical problem solving. *School Science and Mathematics, 117*, 146–157.

Chapter 2
Image Schema Theory

2.1 Introduction

The concept of image schema is an intuitively resonant one, since it appears to naturally explain the cognitive origin of constructs such as the number line, the coordinate system, infinity, and so on (Chap. 1). But what does an image schema actually look like in the mind, so to speak? It can be suggested that it is primarily from an analysis of the features of mathematical diagrams that we can get an indirect (inferential) look into the mathematical mind. The number line, cardinality proofs such as those by Galileo and Cantor, which involve linearity (Chap. 1), and other diagrammatic forms provide elemental visual cues for visualizing the path image schema, given that the diagrams may actually be guided in their physical realization by inner image-schematic structure.

Consider the path schema more specifically. To reiterate here, it derives initially from our sense of movement in physical space, which can be oriented in various directions, such as forward, backward, diagonally, in angular ways, and so on. This sense is then encoded by the brain as a path schema, which manifests itself as a common figural form in verbal metaphorical expressions such as "moving forward in life" and "looking back on life," among a myriad others, and, as discussed, in mathematical concepts that involve linear structure. Language and mathematics, therefore, provide the relevant cues that the schema has some imaginary linear form in the mind and, as such, can be manipulated in various ways to produce different concepts. So, a single horizontal linear form of the schema is implicit in the number line, while horizontal and vertical linear forms are blended in the mind in a specific way to produce new imagistic structure that can be mapped onto a new target domain—the coordinate system, which is itself realized by a physical diagram consisting of a vertically oriented number line crossing with a horizontally oriented one at right angles. Now, in this new conceptual realm, the movement and location of numbers can unfold in different directions, since there are two number paths, a horizontally oriented and a vertically oriented one. The end goal of the movement is the location of points in the plane which can now be represented as ordered pairs of numbers (coordinates) standing for the dimensions of the horizontal and vertical paths (distances) from the starting point (the origin) with relation to the intersecting number lines (axes).

This newly formed conceptual apparatus allows us, now, to think of numbers and of the geometric figures constructed within it in a more sophisticated blended fashion, that is, it becomes itself a conceptual *source* for doing arithmetic, algebra, and geometry in tandem, which constitutes the new *target* domain (analytic geometry). We can now describe a line, a circle, or some other curve in algebraic terms. So, for instance, by plotting the points described by the equation $x^2 + y^2 = 4$, we get a circle whose diameter is four units—two to the left of the origin and two to the right (Fig. 2.1).

At this point, one may ask: Is this nothing more than a creative isomorphism, as discussed in Chap. 1 and therefore not a true psychological explanation of the origins of the coordinate system? But, then, every theory is an isomorphism between facts or observations and a modeling of the facts—an inferential form of thinking

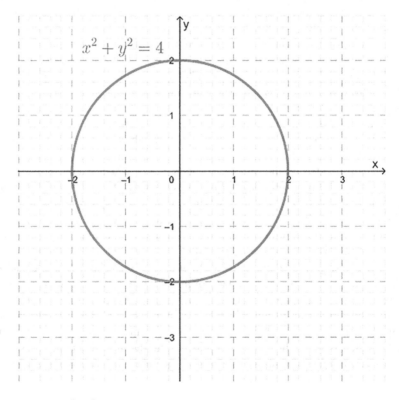

Fig. 2.1 Plotting $x^2 + y^2 = 4$

that we use as an instrument of interpretation, as the philosopher of science, Karl Popper (1934: 280), once insightfully put it:

> Bold ideas, unjustified anticipations, and speculative thought, are our only means for interpreting nature: our only organon, our only instrument, for grasping her. And we must hazard them to win our prize. Those among us who are unwilling to expose their ideas to the hazard of refutation do not take part in the scientific game.

Image schema theory is no different from other theories and, as such, must endure the "hazard of refutation," to use Popper's phrase. Barring a complete refutation, it is therefore to be viewed as plausible as a theoretical construct, providing a concrete framework that can be litmus-tested in three main areas: (1) the learning and teaching of mathematics, (2) the use of diagrams as sources for envisioning image-schematic structure, and (3) the AI modeling of image schemas themselves. All three will be discussed in Chap. 4. In this chapter, the aim is to evaluate image schema theory as a viable construct that can be enlisted to explain, or at least describe, various aspects of math cognition (see Hedblom, 2021 for an overview).

2.2 Image Schemas

Image schemas were introduced into the cognitive sciences originally by George Lakoff and Mark Johnson (Johnson, 1987; Lakoff, 1979, 1987; Lakoff & Johnson, 1980, 1999), who defined them as recurring mental forms extracted by the brain from lived experiences and transformed into patterns of thought that then guide the generation of abstract concepts via metaphorical mappings (discussed in the next chapter). For instance, expressions such as "There is a long way to go in your career," "She is moving forward at breakneck speed," "I like to look back over my life's journey," and others are instantiations of the conceptual metaphor, *life is a journey*—that is, they are specific linguistic metaphorical instantiations of that thought formula, in which *life* is the target domain and *journey* the source domain. The image schema that guides the mapping in this case is the path one, which allows us to think of *life* in terms of a *journey* that is akin to physical motion along an endless path, which does have an end point. The ontological properties of the mapping process involved in producing the *life is a journey* conceptual metaphor can be shown as follows (Fig. 2.2).

As Lakoff (1979: 129) noted before writing *Where Mathematics Comes From* with Núñez, mappings preserve the image-schematic structure that constitutes a source domain as it is projected onto a target domain. So, the physical features of *journeys*, such as traveling from one place to another, perhaps overcoming obstacles on the path toward the destination or goal, or maybe turning back for some reason, are preserved and mapped in various ways onto the target domain of *life*: "Life is a journey through uncharted territory," "Sometimes you must overcome obstacles on your way to your future life," "If you turn back on life, then you will not succeed," and so on. The key aspect in terms of image schema theory is that mappings are

Fig. 2.2 Conceptual metaphor mapping

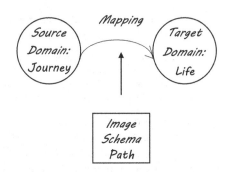

guided or sustained by image schemas or a combination of schemas. Below is a set of conceptual metaphors identified by Lakoff and Núñez (2000) as cases-in-point:

- *Change is motion*, which undergirds the calculus, limits, and other mathematical ideas, is based on path continuity and motion on a continuum, where the path can be linear or curving.
- *Sets are collections of mathematical objects*, which underlies such mathematical concepts and procedures as sets, arithmetical operations, and group theory, is guided by the container schema.
- *Continuity is gapless,* which subtends ideas such as mathematical infinity, is motivated by the schema of an unending continuous path.
- *Functions are sets* (*collections*) *of ordered pairs of objects* (*numbers or other symbols*), which undergirds coordinate and complex number systems, among others, is based on a combination of path, container, orientation, and rotation schemas.
- *Geometric figures are objects in space*, which underlies the representation and mathematical analysis of geometrical objects, is based on a combination of schemas such as spatiality, motion, the cycle, among others;
- *Numbers are object collections*, which is the conceptual basis of arithmetic and set theory, is based on the container schema.
- *Recurrence is circular*, which undergirds such concepts as π, sequences, and series, is based on several schemas including orientation, iteration, center–periphery, and a few more.

Each of these metaphors constitutes the underlying conceptual core of branches and systems such as the calculus (change is motion), infinity (continuity is gapless), set theory (numbers are object collections), and so on. In this framework, metonymy (the part for the whole) can be envisioned as the mechanism that allows for generalizations from particular instances—it is based on a part-whole image-schematic mode of thinking which underlies concepts such as fractions and fractals, among others. As Solomon Marcus (2012: 124) has insightfully observed, even the language invented and used by mathematicians reveals an intuitive awareness of the metaphorical structure of their conceptualizations:

2.2 Image Schemas

For a long time, metaphor was considered incompatible with the requirements of rigor and preciseness of mathematics. This happened because it was seen only as a rhetorical device such as "this girl is a flower." However, the largest part of mathematical terminology is the result of some metaphorical processes, using transfers from ordinary language. Mathematical terms such as *function, union, inclusion, border, frontier, distance, bounded, open, closed, imaginary number, rational/irrational number* are only a few examples in this respect. Similar metaphorical processes take place in the artificial component of the mathematical sign system.

To reiterate, image schemas are formed when the brain extracts meaningful information from our everyday experiences (walking along a path, orienting one's look in various directions, putting things into containers, moving without stopping, etc.). Once inserted in neural circuitry, they guide the formation, understanding, and use of expressive and knowledge systems (Hampe, 2005; Oakley, 2010). They exist in both static and dynamic versions (Cienki, 1995), which constitute what will be called meta-image-schematic structure below.

Consider how a single image schema, the verticality schema, is diffused throughout human cognitive systems. Neurologically, it is formed from the sensation of looking up and down, becoming an ontological guide to the ideation and understanding of various cognitive, emotional, and social phenomena. In verbal discourse, for instance, it manifests itself in the conceptual metaphor, *up is better/down is worse*, as can be seen in typical expressions such as: "I'm feeling up," "They're feeling down," "I'm working my way up the ladder of success," "His status has gone down considerably," and so on. The fact that we interpret these in this way indicates that the schema is unconsciously guiding our understanding of their meaning. This exact same type of metaphorical conceptualization is manifest in musical expression and interpretation, whereby higher tones in a composition are felt to convey an emotional rise, while lower ones are perceived instead to communicate an emotional drop. During oral speech, a raising hand gesture designates notions of amelioration, betterment, growth, etc., accompanying expressions such as "Things are looking up," "You must reach for the sky," and so on, whereas the lowering of the hand designates opposite notions, such as "The world is going down the drain," "You must not lower your expectations," etc. The fact that such gestures are spontaneous and largely unconscious is highly suggestive of the unconscious operation of the verticality schema in vocal communication, both activating the hand movements and providing an ontological mechanism for interpreting them.

This pattern of meaning-making is found across mathematics. Consider the image schema of the container, which, to reiterate, is the ontological mechanism that is behind the sense that numbers are akin to objects in containers, and which is mapped onto arithmetic to produce the *arithmetic is object collection* conceptual metaphor. It crystallizes as a mind-form from activities of putting objects in all kinds of physical containers, including the hands as receptacles, and keeping track of how many there are in them. From this experience, the schema informs thinking in a large number of areas, not only arithmetic, but geometry and other branches as well. It is the basis, for instance, in determining area, volume, and other geometrical measurement procedures. As Postnikoff (2014) aptly puts it, we unconsciously understand how and why such procedures and concepts work because they emanate ultimately from the "human experiences of collections of objects," which "possess extensive image-schematic structure that facilitate our agential interactions with those collections." Because mappings preserve the features of an image schema, the concept *arithmetic is object collection*, once formed in the mind, is able to activate reasoning mechanisms, which are subtended, when unpacked, by the container image schema—hence, the construct of sets, groups, and fields. The point is that, at the most fundamental level of cognition, each one of these can be traced to the container schema, which itself is derived from a particular set of experiences with physical containers—no matter what shape they take or substance they are made of.

There is supporting developmental evidence, as briefly mentioned in Chap. 1, which indicates that ontologically primitive schemas such as the container and path ones emerge spontaneously in childhood. The container schema, for instance, has been found to provide the child with a mental template for differentiating quantitative notions, such as the difference between "bigger" and "smaller" quantities, while the path schema has been found to provide the child with the complementary template for grasping the notion of countability in a linear fashion, as in the number line (Chap. 4). Strong support for the psychological validity of such schemas comes, in fact, from the field of elementary mathematics education, where both schemas are used intuitively by educators in early grades to impart basic notions of arithmetic through derived pedagogical methods and techniques. One of the first educational methods to incorporate image-schematic-based pedagogy intuitively into its teaching practices is the Montessori method (for example, Montessori, 1912). It is "intuitive" because it was founded before the advent of image schema theory. The container and path schemas are found inherently throughout the method. A few examples are as follows:

- *Container schema*: In the method, the container schema is the basis of a large array of pedagogical techniques. One of these involves the use of beads and trays for teaching children to add small numbers. For example, in one application, numeral cards are placed before the trays, and the child is asked to put the proper quantity of unit beads into the tray with that number. The trays are also used for teaching basic arithmetical operations, such as uniting the quantities of two trays into a third tray to impart the notion of addition.
- *Path schema*: The path schema is also the source of various arithmetical teaching techniques. For example, beads are employed to get children to physically see and touch numbers and to allow them to see the difference between two beads laid out in sequence versus, say, seven beads, which is intended in part to get children to conceptualize the importance of number layouts such as the number line.

It is relevant to note that some of the main conceptual metaphors identified by Lakoff and Núñez are grounded on these two primitive image schemas—for instance, *arithmetic is object construction, arithmetic is the use of a measuring stick, arithmetic is motion along a path,* and *classes are containers.* The *object*

2.2 Image Schemas

construction metaphor reflects the imaginary action of taking objects from a container or several containers to construct new mathematical structures; when blended with the part-whole metonymic schema, it undergirds notions such as fractions and fractals; the *arithmetic is motion along a path*, as discussed, is fundamental to the ideation and interpretation of notions such as the number line, infinity, and the like; the *measuring stick* metaphor is behind activities such as laying out quantities or symbols in a linear (including a diagonal) fashion in order to see how many there are or how they relate with each other, which is the basis of many mathematical ideas and proofs, such as the diagonal proof used by Cantor and Gödel; and the *classes are containers* metaphor is an extension of the container schema generalized to set theory and other areas such as fields, groups, etc.

As the mind-landscape of mathematics becomes populated with image schemas layered upon other (Chap. 3), similarities among them can be extracted to envision or even create new ideas. So, image-schematic notions of numbers as points laid out in geometric configurations can be seen as suggestive of an intrinsic cognitive relationship between arithmetic and geometry. This was arguably the cognitive basis for the Pythagorean idea of figurate numbers—numbers that can be displayed geometrically. For example, square integers, such as 1^2 (= 1), 2^2 (= 4), 3^2 (= 9), 4^2 (= 16), and 5^2 (= 25), can be displayed with square arrangements of objects, such as dots, forming a one-to-one correspondence (Fig. 2.3).

By portraying square numbers in this way, the Pythagoreans established one of the first principles for what came eventually to be called number theory, discovering that they are equal to the sum of consecutive odd integers:

$$1^2 = 1 = 1$$
$$2^2 = 4 = 1+3$$
$$3^2 = 9 = 1+3+5$$
$$4^2 = 16 = 1+3+5+7$$
$$5^2 = 25 = 1+3+5+7+9$$
$$6^2 = 36 = 1+3+5+7+9+11$$
$$\ldots$$
$$n^2 = 1+3+5+7+9+11+\ldots$$

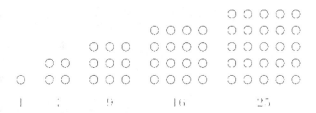

Fig. 2.3 Figural composition of square numbers

This results from the fact that, to form each new square figure, a successive odd number of dots to the preceding figure must be added—that is, to the square figure for the number 1, three new dots must be added to produce the next figure for 4; then to the square for the number 4, five more dots must be added to produce the next figure for 9; and so on. The figurate square number of $6^2 = 36$ can be shown with consecutive additions of dots partitioned off to show the additions (Fig. 2.4).

This display now shows visually that 36 is the sum of the first six odd numbers, since it is made up successively of $1 + 3 + 5 + 7 + 9 + 11$ consecutive partitions of dots. The Pythagoreans called the odd numbers *gnomic* because the partitioning method above resembles a carpenter's square known as a *gnomon*, showing the patterned arrangements of the dots.

As another example, consider the Pythagorean notion of triangular numbers, in which the first triangular number, 1, consists of 1 object; the second triangular number, 3, consists of $1 + 2$ objects; the third triangular number, 6, consists of $1 + 2 + 3$ objects; and so on. Each successive triangular number is thus obtained by adding a row of objects (such as dots) to the bottom row, which will have one more object than the corresponding row in the previous triangular number (Fig. 2.5).

This layout led to the discovery that the nth triangular number is the sum of the first n counting numbers:

1st triangular number : $1 = 1$

2nd triangular number : $3 = 1 + 2$

3rd triangular number : $6 = 1 + 2 + 3$

4th triangular number : $10 = 1 + 2 + 3 + 4$

5th triangular number : $15 = 1 + 2 + 3 + 4 + 5$

6th triangular number : $21 = 1 + 2 + 3 + 4 + 5 + 6$

...

n^{th} triangular number : $1 + 2 + 3 + ... + n$

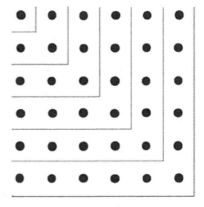

Fig. 2.4 Figural composition of 6^2

2.2 Image Schemas

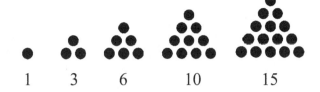

Fig. 2.5 Figural composition of triangular numbers

The reason for this is as follows. To produce the second triangular number, two dots must be added to the configuration; to produce the third one, three dots must be added; to produce the fourth one, four dots must be added; and so on. The point to be emphasized here is that these diagrams show an image-schematic structure based on the layout of objects in a combinatory fashion based on a sequential patterning that reflects a blend of layout and counting conceptual metaphors, traceable to a blend of path and container schemas. The psychological validity of this analysis can be seen, anecdotally, in how these patterns are used in mathematics education fruitfully. For example, although they do not make explicit reference to image schema theory, Petersson and Weldemariam (2022) intuitively used it as a pedagogical frame to impart the sense of the numbers and shapes as intertwined to preschool learners. The researchers used a game as a pedagogically focused activity whereby the children were tasked with parking toy cars in a rectangular shape, so as to determine for which number of cars this was possible. Thirteen preschool children were divided into three separate age groups, from 4 to 6 years, together with their teacher, who participated in the study. The results were significant, as the teacher's observations also indicated, given that the game was not only engaging for the children but also highly effective in activating mathematical thinking in them. As the researchers conclude (Petersson & Weldemariam, 2022: 715): "The idea of figurate numbers has proven useful not only within the history of mathematics but also in present mathematics education because it provides a concrete and multi-modal way to represent both numbers and several important mathematical ideas."

As the Pythagorean discoveries show, primitive image-schematic thinking is operative in laying the foundations for mathematical discovery, even though it is not consciously realized until diagrams, such as those above, are created which put such thinking on display in concrete visual ways—as will be discussed further in Chap. 4. There may, of course, be other ways of explaining the evolution of such ideas, but the point here is that image schema theory provides a plausible analytical tool for deconstructing how the ideas may have come about. Primitive image schemas emerge, as mentioned, in childhood thinking patterns (Hedblom, 2021) during the secondary "math" phase. Conversely, image-schematic thinking based on a cascade of mappings (Chap. 3) arises through effort and instruction—which occurs at the levels of "math" and "mathematics." From research such as that of Núñez et al. (2012), it can be deduced that the "mathematics" level of understanding based on advanced mapping processes requires explicit training to acquire (in most people).

2.3 A Typology

The starting point for image schema theory is not Lakoff and Núnez's *Where Mathematics Comes From*, as is sometimes assumed within the math cognition field, but can be traced, rather, to the work of linguist Ronald Langacker (1987, 1990, 1999), one of the first promoters of the cognitive stream within linguistics in the 1980s and 1990s, based largely on conceptual metaphor theory. Langacker argued, for instance, that the parts of speech themselves were grounded on image-schematic thought. Nouns, he claimed, invariably encode the image schema of a region, which is a reification of the container schema, in bounded (closed) or unbounded (open) mental versions. This is why a count noun such as *leaf* is envisioned as eliciting a conceptually bounded referential region and a mass noun such as *rice* as an unbounded one. Now, this difference in container structure is the source of various differential grammatical features in English. Because bounded referents can be counted within a conceptual container, a noun such as *leaf* has a corresponding plural form (*leaves*), whereas a noun such as *rice* does not, since it enfolds an unbounded region in the mind and is thus perceived as uncountable. The noun *leaf* can be preceded by an indefinite article (*a leaf*), *rice* cannot, since the indefinite article is specific as to quantitative reference. The set of grammatical and semantic features that are intertwined with the image-schematic entailments of specific linguistic forms is rather large and need not concern us here.

At around the same time, Mark Johnson (1987) and George Lakoff (1987) were extending Langacker's model to enfold all kinds of conceptual phenomena, not just semantic and grammatical ones, although Lakoff (1979) had already laid the groundwork for a proto-image schema theory, as we now understand it, even before Langacker. Johnson (1987) provided one of the first concrete typologies of image schemas that, as elaborated in *Where Mathematics Comes Form* by Lakoff and Núñez, can be enlisted to explicate the potential source domains for different mathematical ideas—note that this is a minimal selection used simply for illustration (see Hedblom, 2021 for an elaborate explication):

- *Container*: As discussed, this schema guides the perception of numbers as objects contained in receptacles of various kinds, and it is the source of concepts such as area, volume, sets, fields, groups, among many others.
- *Path*: Also as discussed, this schema allows for a conceptualization of numbers as objects laid out according to some figural structure—horizontal, vertical, logarithmic, etc.; it is thus the source of constructs and ideas such as the number line, the BMI, etc. This schema enfolds various subtypes including specific configurations of objects (such as figurate numbers).
- *Cycle*: This schema is the source of a whole array of concepts that involve circularity or curvature more generally, including π, topological equivalencies (circles as types of ellipses), etc. It is also related to the schemas of *iteration*, which is, for instance, the source of the sequence notion, and *rotation*, which subtends various areas of geometry, trigonometry, topology, vector analysis, complex numbers, and so on.

2.3 A Typology

- *Force*: This schema is the source of various concepts that involve, metaphorically, "forcing" some mathematical object out of some configuration in order to identify its properties or to indicate what it entails. The Gödelian proof that there exists a proposition in a system of propositions that cannot be proved or disproved involved forcing it conceptually out of the system for analysis. This schema includes the following elements: a source and target of the force; a direction and intensity of the force; a path of motion of the source; and a sequence of causation. It is also the inspiration of many ideas in mathematical physics such as the relation between force and mass.
- *Counterforce*: This is the counterpart to the *force* schema, alluding to a conceptual force that countervails or opposes an existing force. The notion of a counterexample in mathematical proofs can be seen to be a counterforce schema. It is thus the basis of the *reductio ad absurdum* method of proof, also known as proof by contradiction. Many classic proofs in mathematics are based on this schema.
- *Orientation*: In tandem with other image schemas, this is behind the ideation of coordinate systems, graphs, sequential patterns, and so on; the orientation is not restricted to the spatial features of the plane or the sphere; it is also the intellectual source of n-dimensional space beyond three-dimensions. The schema enfolds various concrete subtypes, including verticality and horizontality schemas, which are blended in the ideation of systems such as coordinate geometry.
- *Compulsion:* This is a subtype of the force schema, derived from the sensation of pushing objects. It is the likely source of various notions in physics and mathematics, including vector theory and the metaphorical notion of eliminating anomalies in sets and other mathematical structures.
- *Blockage*: This is another subtype of the force schema in which motion is stopped or redirected by the presence of an obstacle. It can be seen to provide an imagistic scenario whereby a proof of a conjecture is "blocked mentally" by either the absence of required mathematics or the presence of unknown factors that inhibit the proof from being devised.
- *Restraint-Removal*: This is a counterpart to the *blockage* schema, involving a removal of a conceptual blockage, based on some new metaphorical mapping or blending of image schemas, which leads finally to a solution or proof of some previous conjecture. The solution of the four-color conjecture with the use of computers is a classic example of the operation of this schema, based on the removal of the available proof methods at the time, and the invention of a new form of proof (Chap. 4).
- *Link:* This schema consists of a blend of two or more schemas, connected ontologically or structurally in some way; it is the likely source of conceptual metaphors such as the BMI.
- *Scale*: This schema is derived from observations of increasing or decreasing scales of measurement or quantity. It is the likely basis of several arithmetical and geometrical computations, including size comparisons (such as equality, inequality, approximation).
- *Balance*: This schema provides mental portraiture of the notion of equilibrium, constituting a source of ideas and constructs such as equations and functions.

Balance scales are used in mathematics education to concretely demonstrate notions such as that of an equation to learners (Otten et al., 2019).
- *Part-whole*: As discussed, this metonymic schema is the likely source of concepts such as fractions and fractals, among various others.
- *Center–periphery*. This schema has three elements: an entity, a center, and a periphery. The center is essential, the periphery less so. It is the core of many ideas in geometry and topology. As a study by Williams (2010) found, it is even operative in how we read traditional clocks.

In a key (2014) article, Mandler and Canovas provided a developmental timetable for the emergence image schemas in childhood, arguing that, before an image schema crystallizes in the brain, there are "spatial primitives" within the neural substrate which are formed during a sensory ("pre-math") stage. These then become the sources of a broader image-schematic cognition based on the observation and experiences related to physical containers, paths, object placement, etc., which then evolve into actual image schemas during the "math" phase of development. Finally, there are "schematic integrations," which correspond to what has been called the abstract "mathematics" phase here. Now, taking some liberty with the Mandler-Canovas typology, spatial primitives can be called "primitive *Gestalten*," defined as "pre-math" mental structures, which take shape in the brain from the experience of things, becoming the basis of image-schematic cognition, as the brain refines the Gestalten into mnemonically retrievable forms embedded in neural circuitry. Finally, this conceptual apparatus becomes the substratum on which abstract mathematical forms and systems result through advanced mapping processes. Below is a highly selective typology of basic image schemas as they relate to mathematical mappings. Note also that some schematic forms are included in different categories, showing that there is some overlap among them (Table 2.1).

A possible psychological critique of a typology such as the one above is that a notion such as "spatial primitives" may be nothing more than a synonym for *percepts*, the traditional notion in psychology of an organized mind-form that results when the brain translates sensory-experiential sensation into structures of memorable thought. Percepts do indeed constitute interpretive filters of incoming information extracting from it what is meaningful in a given context, a process that is largely unconscious. However, it can be argued that percepts are not coincident semantically with spatial primitives, which are integrations of sensory information, not just perceptions of it, which create cognitive frames that allow us to interpret the world in specific ways. Once formed, these become the source of detecting recurring patterns of quantity and spatiality in the information that comes from experience.

2.4 The Gestalt Background

Table 2.1 A selective image schema typology

Gestalten	Image-schematic forms	Math forms (systems)
Spatiality	Path, orientation, verticality, horizontality, configuration, etc., including up–down, front–back, left–right, near–far, object contact, straightness, curvature, cycle	Number lines, coordinate systems, geometric figures, graph theory, trigonometry, figurate numbers, infinity, countability, etc.
Containment	Container, center–periphery, in–out figural events, surface versus inner, full–empty, boundaries, closed, open	Arithmetical operations, set theory, decidability (including P = NP hypothesis), etc.
Motion	Source–path–goal, movement along paths, end goal, endless movement	Limits, calculus, continuous versus discrete mathematical systems, etc.
Balance	Equilibrium, evenness, congruity of forms, equality of size, or quantity	Equations, functions, proofs of equality, etc.
Force	Compulsion, blockage, counterforce, enablement, attraction, resistance, causality	Vector arithmetic, abstract algebra, relations among equations and functions, proofs such as *reductio ad absurdum*, etc.
Unity-multiplicity	Part-whole (metonymy), uniting, splitting, constructing, deconstructing	Fractions, fractals, set-theoretic notions, etc.
Existentiality	Equality, matching, comparing, subitizing	Existence theorems, decidability, equations, inequalities, etc.
Scale	Path, measurement, equality, range, size	Notions based on measurement systems, including scalar versus vectorial notions

2.4 The Gestalt Background

The notion of image schema dovetails considerably with the notion of *Gestalten* (mind forms) associated with the Gestalt school of psychology (Koffka, 1921; Köhler, 1925; Wertheimer, 1923) and especially with principles such as *closure, proximity, continuity, similarity, constancy,* and *figure-ground*. Closure refers to the tendency of the brain to perceive things as complete and unified, "filling in" missing parts to produce a unitary form—a tendency that prefigures the image schema of the perception of how containment occurs. The figures below are interpreted as a circle and a square, even though their boundaries (perimeters) are not continuous (Note that the figures in the sections below are commonly used ones within classic or standard Gestalt theory writings, as for example, Koffka, 1935; Hamlyn, 1957; Murray, 1995) (Fig. 2.6).

Proximity refers to the phenomenon of perceiving objects that are close to each other as belonging together as categories with the same characteristics or properties.

Fig. 2.6 Closure (Wikimedia Commons)

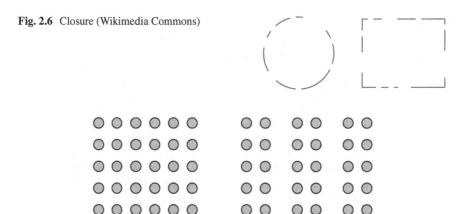

Fig. 2.7 Proximity (Wikimedia Commons)

Consider the two configurations below, (a) and (b), which contain the same number of dots of uniform size and shape (Fig. 2.7).

Even though the dots are exactly the same, we tend to interpret (a) as one group of the dots and (b) as three groups of the same dots because of the different arrangement of the dots. In (a) they are equally spaced apart suggesting the outline of a square figure and thus as constituents of the square, while in (b) the two vertical spaces between the dots suggest the outline of three columns, each of which constitutes a separate grouping of the dots. This principle prefigures the linkage image schema, among others, which, as discussed, envisages two or more entities as connected physically or metaphorically, identifying some ontological bond between them.

Continuity refers to the tendency of the brain to perceive certain configurations as arrangements of discrete objects constituting a continuous form, rather than disjointed or discontinuous forms (Fig. 2.8).

In the figure above, we are more likely to perceive two overlapping curves of dots, rather than four curves meeting in a center. Image schemas based on some form of continuum figuration are based on this principle.

Similarity refers to the perception that certain things share common properties, which is why they are grouped together (Fig. 2.9).

When looking at the array of dots in the figure above, we tend to perceive alternating rows of similar shades—darker and lighter—as belonging to the same category according to shade. The figure is thus perceived as enfolding two kinds of interlacing categories. This is the likely source of image schemas involving groupings based on similarity and equality that are found throughout mathematics.

Constancy is the perception of inherent connections in specific types of objects. An example is a mosaic, which is made up of tiles that are put together in a collage

2.4 The Gestalt Background

Fig. 2.8 Continuity (Wikimedia Commons)

Fig. 2.9 Similarity (Wikimedia Commons)

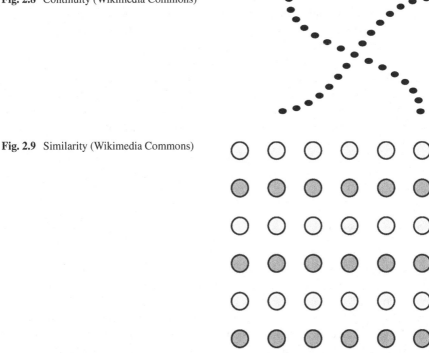

form to make a new, unified whole. This overlaps with several of the previous perceptual processes, indicating a tendency of the brain to seek pattern and, where there might be none, to impose it. Constancy underlies the image-schematic conception of numbers and spaces as sharing properties on the number line, whereby the space between two consecutive numbers represents a measurable length and the number of these lengths constitutes various schematic concepts, from continuity to measurability.

The figure-ground principle implies that we instinctively perceive objects as either being in the foreground or the background. As Frisch and Julesz (1966) found in their research, this principle is a primary one in geometry, which they describe as follows (Frisch & Julesz, 1966: 389)—note that this principle prefigures image schemas such as the center–periphery one:

> Constructs of random geometry were applied to the problem of figure-ground perception. Random-dot images of black and white dots with various area fractions and tessellations (square and triangular lattices) were used as stimuli…It was first conjectured and then experimentally verified that figure-ground perception is not affected by the various tessellations used. Thus, figure-ground phenomena depend only on the area fraction of the white and black dots in the stimulus. It was also experimentally shown that size-constancy prevails in figure-ground perception, but brightness-constancy does not.

The overriding premise of Gestalt psychology is that the brain avoids randomness or incompleteness, seeking pattern and completeness instead. The formation of mental Gestalten can be said to be based on spatial primitives derived from the human sensory apparatus, which the brain then transforms into image schemas. Infants live in a mind-state sometimes called a sensorium (Carpenter & McLuhan, 1960; Howes, 1991), that is, they process information through their sensory system, which leads to the creation of primitive sense-based models of the world in the brain. However, in a remarkably short period of time, they start replacing sense-based understanding with conceptual knowledge. This event is extraordinary—all that children require for the conceptual mode of knowing to be set in motion is simple exposure to concepts in social context through language, pictures, simple arithmetic, and other kinds of symbol-based systems of representation and communication. From that point on, children require their sensory apparatus less and less, becoming more and more dependent on their conceptual mode.

2.5 Stasis–Motion–Force

As mentioned, Cienki (1995) introduced a basic ontological dichotomy in image schema theory—static versus dynamic schemas. These are overarching or "meta-image-schemas" on which others are built. The static one can be named the *stasis* meta-schema and the dynamic one, which is its counterpart, the *motion* meta-schema. The former would enfold schemas such as containment, blockage, and the like, while the latter would subsume schemas such as the path and movement ones. A *force* meta-schema (Talmy, 1981, 1988) can be added to Cienki's classification, which subsumes schemas involving motion from one state to another as caused by some force, physical or imaginary. This triadic meta-image-schematic system undergirds the array of derived image schemas that inform source domains that are then mapped onto target domains to produce complex ideas in mathematics. To see what this implies, consider two classic episodes in the history of mathematics—Zeno's paradoxes and Euler's graph theory.

As is widely known, the pre-Socratic philosopher Zeno of Elea formulated a series of paradoxes that challenged logic and mathematics during his times (McGreal, 2000). Inherent in several of these is the path image schema, as can be seen, for instance, in his *Achilles and the Tortoise* paradox, recounted by Aristotle in his *Physics* (350 BCE). Below is a paraphrase:

> Achilles decides to race against a tortoise. To make the race fairer, he allows the tortoise to start at half the distance away from the finish line. In this way, Achilles will never surpass the tortoise. Why?

The well-known argument goes as follows. In order for Achilles to surpass the tortoise, he must first reach the halfway point, which is the tortoise's starting point. But when he does, the tortoise will have moved forward a bit. Achilles must then reach this new point before attempting to surpass the tortoise. When he does, however, the tortoise has again moved a little bit forward, which Achilles must also reach again, and so on ad infinitum. In other words, although the distances between

2.5 Stasis–Motion–Force

Achilles and the tortoise become smaller, infinitesimally so, Achilles will never surpass the tortoise. Of course, in reality Achilles will do so, because motion is continuous, not a series of stop-and-go movements. Without going into the philosophical-mathematical debates that the paradox engendered, the point here is that it exemplifies, in miniature, how the stasis meta-schema guides the enigmatic argument—the linear path in which Achilles and the tortoise are carrying out the race is portrayed by Zeno as consisting of static (stopping) points. Reaching any of these points by Achilles involves, of course, the motion meta-schema, which is mapped onto the linear path as decreasing distances between static points, rather than as a continuous advance. Now, a cognitive resolution of the paradox can be seen if the force meta-image-schema is mapped onto the scene, because it would involve adding to it a force of continuous motion (literally) rather than stoppage at points on the path. One could claim, arguably, that Zeno's omission of this schema in the description of the scene is what produces the paradox. The fact that the distances between Achilles and the tortoise become infinitely smaller in a patterned continuous way as Achilles approaches the tortoise is what likely led to the theory of limits and then of the calculus (Boyer, 1959: 295).

Another of Zeno's paradoxes, the *Race Course Paradox* (also called the *Dichotomy Paradox*) shows the same type of manipulation of meta-image-schematic thinking. The paradox states that a runner will never reach the end point of a linear race course because, before getting there, the runner must reach the halfway point first and then another half of the remaining distance (= a quarter of the way), after which the runner must cover yet another half distance (= one-eighth of the way), and so on. So, the runner will never get to the finish line, even though we know in reality that the runner actually does. This paradox, like the previous one, treats movement as segmentable into an infinity of static points, exclusive of the force meta-schema.

The discovery of graph theory by Euler is another significant event in the history of mathematics that illustrates, in microcosm, how meta-image-schematic cognition undergirded its inspiration. The theory is traced to a famous 1736 paper, which Euler presented to the Academy in St. Petersburg, and which he published subsequently in 1741. The paper describes an actual layout of bridges in the town of Königsberg through which the Pregel River runs. In the river area there are two islands connected with the mainland and with each other by seven bridges. Below is a sketch showing the location of the seven bridges in the landscape (Freiberger, 2016) (Fig. 2.10).

The residents of the town would often debate whether or not they could cross each bridge once and only once, returning to the starting point. They found that they could not do it without doubling back on one of the bridges. No one could explain why it thus seemed to be impossible to traverse the bridges with a unique route. Euler became intrigued by the conundrum—a conundrum clearly based on the path schema, and on the relation between static points in the bridge network and on the directional motion between the points in the bridge layout, now called a network.

The gist of Euler's well-known classic solution to the dilemma can be encapsulated as follows: First, to eliminate the irrelevant aspects of the Königsberg scene,

Fig. 2.10 The Königsberg layout (Wikimedia Commons)

Fig. 2.11 A graph of the Königsberg network

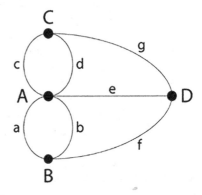

retaining its essential elements, we draw a graph of the area, representing the land regions with capital letters (A, B, C, D) and the bridges with small letters (a, b, c, d, e, f, g). The capital letters stand for the points (or vertices) in the network and the small ones for the paths in it. These can be joined as shown (Mathematics Stack Exchange, 2015) (Fig. 2.11).

Now, putting a writing instrument such as a pencil on any point and attempting to traverse the entire network without doubling back at some point is impossible. The reason is that a network can have any number of even paths in it, because all the paths that converge at an even vertex are "used up" without having to double back on any one of them. For example, at a vertex with just two paths, one path is used to get to the vertex and another one to leave it. Both paths are thus used up without having to double back over either one of them. In a four-path network, when we get to a vertex we can exit via a second path. Then, a third path brings us to the other vertex, and a fourth one gets us out. All paths are once again used up, and there was no doubling back on any of them. At an odd vertex, however, there will always be one path that is not used up. For example, at a vertex with three paths, one path is used to get to the vertex and another one to leave it. The third path can take us back to the vertex. But to get out, we must double back over one of the three paths.

In the Königsberg network each vertex is odd, consisting of three paths. It is thus a network with odd vertices and thus not traversable without doubling back. As such, Euler's graph is a diagram that is designed on the underlying image-schematic structure of the scene, including the meta-image-schema of stasis (which underlies the concept of vertices as stopping points), motion (going and exiting the vertices via specific paths or edges), and force (making or not making one's way through the network without doubling back). This system is a dynamic one and prefigures the physical structure of electric circuits and neural networks in AI. Of course, this does not explain how Euler used his background knowledge and his imagination to come up with the solution. That is a question for cognitive psychology generally, not image schema theory specifically. The point here is that it likely entailed an activation of meta-imagistic thinking that Euler projected onto a simple diagram that captured the essential mathematical structure of the network.

Now, once incorporated into "mathematics," Euler's solution became the basis for graph theory, which, as a conceptual metaphor, could then be mapped onto other domains of mathematics to create other branches and to show connectivity among them. As Paoletti (2011) aptly observes:

> After Euler's discovery (or invention, depending on how the reader looks at it), graph theory boomed with major contributions made by great mathematicians like Augustin Cauchy, William Hamilton, Arthur Cayley, Gustav Kirchhoff, and George Polya…Other famous graph theory problems include finding a way to escape from a maze or labyrinth, or finding the order of moves with a knight on a chess board such that each square is landed on only once and the knight returns to the space on which he begun.

2.6 System Thinking

Graph theory shaped the future course of mathematical thinking in many ways, congealing into what can be called "system thinking," a way of understanding and using the interrelated elements of the theory as part of a unified whole and projecting these, via mappings, onto other domains (Chap. 3). This process characterizes both institutional thinking—"mathematics" as a discipline—and the ability of individuals, via training and continued exposure to the system, to use it for various subjective purposes. This notion is based partly on Russian psychologist Lev Vygotsky's (1962) idea of acquired knowledge as an "organizing system" of the concepts that originate and develop within us at specific points in time. A branch of mathematics, such as graph theory, evolves gradually into an organizing system of concepts that allows us to use it for further exploration within and outside of mathematics. In the field of artificial neural networks, as briefly mentioned, graph theory is the basis for designing network topologies and routing algorithms, as well as for optimizing the transmission of data and helping determine the most efficient paths for data packets to travel from source to destination.

The point is that a new branch of mathematics, and the system thinking that it engenders, does not emerge in a vacuum; it is the outcome of enlisting image-schematic thinking creatively, allowing the mathematician to turn a practical situation into a diagrammatic one (at least in the mind, if not on paper). The diagram is, in its essence, an image-schematic external model of a situation, which then provides insights into the mathematical structure of the situation itself, from which mathematical properties can be extracted, and which, in many cases, lead to a new branch and new form of system thinking. In the case of the Königsberg scenario, the image-schematic picture of a network of paths in which points are connected by the paths presents the situation in outline (schematic) form from which structural properties can be extracted. The whole network can be seen to be subtended cognitively by the three meta-image-schemas of stasis (the locations of the points and paths), motion (movement along the paths), and force (the direction of the movement and its possible outcomes). The whole process can be represented as follows (Fig. 2.12).

In this framework, image schemas constitute the sources of the relevant insight, which when applied to a situation lead to the crystallization of a mental-diagrammatic model of the situation. After the model is established as bearing general implications, it is mapped over and over onto other domains to produce system thinking within a large portion of the entire field ("mathematics").

Consider the introduction of set theory into "mathematics" as another case-in-point. This would hardly have come into the mathematical mind without the insights that were prompted by two image schemas—the path schema which was mapped onto the number line concept by Wallis and the container image schema, which allowed for the imaginative manipulation of symbols (numbers) as displaying differential properties in terms of the contents of different containers. As discussed, Cantor's demonstrations of cardinality in the 1870s were ultimately implanted in Wallis's conceptual metaphor of the number line, first utilized in this way by Galileo (as discussed).

What is remarkable is that set theory became, shortly after Cantor, a fundamental mode of system thinking within mathematics. The implications of this form of thinking for the concept of infinity and its multiple applications became the source of further cogitation based on subsequent mappings. This is indicated by the fact that Cantor (1891) then went on to prove something that was mind-boggling at the time—namely that there is an infinite set of cardinalities and thus different orders of infinity. Cantor's widely known proof is remarkable for its simplicity. Suppose we

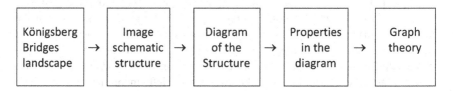

Fig. 2.12 The evolution of graph theory

2.6 System Thinking

take all the possible decimal numbers that exist between 0 and 1 on the number line, labeling each number $\{N_1, N_2, N_3...\}$. The numbers below are just a sampling:

$$N_1 = .4225896...$$
$$N_2 = .7166932...$$
$$N_3 = .7796419...$$
$$...$$

Is it possible to construct a number that is not on the line between 0 and 1? Let us call it C. To create it, we do the following: (1) for its first digit after the decimal point, we choose a number that is greater by one than the first digit in the first place of N_1; (2) for its second digit, we choose a number that is also greater by one than the second number in the second place of N_2; (3) for its third digit, we choose a number that is similarly greater by one than the third number in the third place of N_3; and so on:

$N_1 = .\underline{4}225896...$

The constructed number, C, would start with 5 rather than 4 after the decimal:

$C = .5...$

$N_2 = .7\underline{1}66932...$

The constructed number would have 2 rather than 1:

$C = .52...$

$N_3 = .77\underline{9}6419...$

The constructed number would have 0 rather than 9:

$C = .520...$

...

Now, the number $C = .520...$ is different from $N_1, N_2, N_3, ...$ because its first digit is different from the first digit in N_1; its second digit is different from the second digit in N_2; its third digit is different from the third digit in N_3, and so on ad infinitum. We have in fact just constructed a different number that appears nowhere in the numerical layouts such as the one above. This is truly an astonishing result, since it implies that there are different orders of infinity.

What the foregoing discussion suggests, overall, is that image-schematic thinking is adaptive to new needs, subserving them in specific ways, producing conceptualizations of situations that are converted into new ideas. Over time, new ideas and forms of system thinking emerge from a cascade of mappings of image schemas (Chap. 3). For example, the notion of modular arithmetic, developed by Carl Friedrich Gauss in his *Disquisitiones Arithmeticae* (Gauss, 1801), is an offshoot of the cyclic image schema, whereby numbers are envisioned as "wrapping around" some circular object, reaching a certain value (the modulus) at some point (the stasis and motion schemas working in tandem). Another example in which image-schematic cognition has played a role in discovery and subsequent system thinking is the incorporation of binary numbers into "mathematics"—a system that involves the metaphorical concept of *full-empty* based on the container image schema, where the digit "1" stands for a "full" container and "0" an empty container. It was German mathematician Gottfried Wilhelm Leibniz who elaborated the first binary system in Leibniz (1703). But before Leibniz, the Chinese had come up with a binary system of their own, called the *I Ching*, devised during the Shang dynasty of ancient China (c. 1766–1027 BCE). It was constructed with two lines—a *yin* (broken line) and a *yang* (unbroken line). The lines were converted into numbers and other symbols. Leibniz was apparently inspired by the *I Ching* to put forth his system of binary arithmetic, publishing it in a 1703 paper titled "Explication de l'arithmétique binaire" ("Explanation of Binary Arithmetic"). In a similar explanatory vein, Grice et al. (2024) suggest that the spark for mathematical discoveries, such as graph theory, comes from "preverbal psychological intuitions or principles of perceptual organization," which produce such systems as algebraic structure that "may be inherent in the representations of the world formed by our perceptual system."

2.7 The Etymology of Schema

The notion of *schema* emerges in antiquity as a means to generalize (schematize) logical ideas. In ancient Greek, it meant "form" or "figure." An equation is such a form. The Pythagorean equation $c^2 = a^2 + b^2$, for example, is a schema for saying the same thing as the sentence "the square on the hypotenuse is equal to the sum of the squares on the other two sides." As such, it turns the information and ideas expressed in the words into a schematic form. In so doing, it takes the semantics in the linguistic sentence out, leaving only the structural (logical) outline of the information. It is this feature of the equation that makes it cognitively powerful for mathematics, since we can now find many more meanings and applications for it, in addition to the original geometrical one. For example, one can now ask what sets of three integers fit the equation and, further, if there are other exponents for which the equation holds in general, $c^n = a^n + b^n$. From the latter deliberation has come much subsequent mathematical contemplation leading to such intriguing ideas as Fermat's last theorem. In other words, schematic notations, such as this equation, suggest meanings would be hard otherwise to contemplate.

Immanuel Kant (1781) also used the term *schema*, as a representation of a general procedure of the imagination by which an image is mapped onto a concept. This was the likely source of the notion of *schema* in psychology, introduced formally into the field by Frederic Bartlett in 1932, defining it as a mental construct that allows for the adaptation of new information into existing conceptual systems. This was adopted a little later by Jean Piaget (1952, 1969) to describe how children develop the ability of anticipating events in the future and making plans for them in terms of what he called *assimilation* and *accommodation*. The former refers to how new information is incorporated by the mind into pre-existing schemas, in line with Bartlett's definition; the latter refers to how existing schemas may be altered or new ones formed by novel information, which might disrupt the structure of pre-existing schemas and lead to the formation of new schema structures. The term *schema* was subsequently extended into anthropology (Nishida, 1999) as a structure for adapting cultural knowledge to the understanding of social situations. When we interact with members of the same society in certain contexts and talk about certain topics with them, cultural schemas are involved in guiding conversations and mutual understanding. These coalesce from a society's collective knowledge (Malcolm & Sharafian, 2002), allowing people to interpret the world in specific cultural ways.

It is easy to see the ontological link between the etymology of the notion of schema and that of image schemas, which appear in the late 1980s within linguistics, as mentioned. The image schema notion, however, came forward to designate something more fundamental—namely, development of a mental form derived from experience that becomes embedded in the brain and fine-tuned for interpreting the world. The image schema of a path, as discussed several times, derives from the experience of walking (as Wallis described his number line) which is then mapped continually onto various domains of mathematical cognition, providing insights that lead to discoveries, such as graph theory. To cite Oakley (2010: 214), image schemas are neither images nor schemas, in the traditional senses of these terms; they are "distillers of spatial and temporal experiences. These distilled experiences, in turn, are what cognitive linguistics regards as the basis for organizing knowledge and reasoning about the world."

References

Aristotle (350 BCE). *Physics*. Internet Archive. https://archive.org/details/aristotlephysics0000tran
Bartlett, F. C. (1932). *Remembering: A study in experimental and social psychology*. Cambridge University Press.
Boyer, C. (1959). *The history of the calculus and its conceptual development*. Dover Publications.
Cantor, G. (1891). Ueber eine elementare Frage der Mannigfaltigkeitslehre. *Jahresbericht der Deutschen Mathematiker-Vereinigung, 1*, 75–78.
Carpenter, E., & McLuhan, M. (Eds.). (1960). *Explorations in communication*. Beacon Press.
Cienki, A. (1995). Some properties and groupings of image schemas. In M. H. Vespoor, K. D. Lee, & E. Sweetser (Eds.), *Lexical and syntactical constructions and the construction of meaning*. John Benjamins.
Euler, L. (1741). Solutio problematis ad geometriam situs pertinentis. *Commentarii Academiae Scientiarum Petropolitanae, 8*, 128–140.
Freiberger, M. (2016). The bridges of Königsberg. *Plus*. https://plus.maths.org/content/bridges-k-nigsberg
Frisch, H. L., & Julesz, B. (1966). Figure-ground perception and random geometry. *Perception & Psychophysics, 1*, 389–398.
Gauss, F. (1801). *Disquisitiones arithmeticae*. Gerhard Fleischer.
Grice, M., Kemp, S., Morton, N. J., & Grace, R. C. (2024). The psychological scaffolding of arithmetic. *Psychological Review, 131*, 494–522.

Hamlyn, D. W. (1957). *The psychology of perception: A philosophical examination of Gestalt theory and derivative theories of perception.* Routledge.

Hampe, B. (2005). *From perception to meaning: Image schemas in cognitive linguistics.* Mouton de Gruyter.

Hedblom, M. M. (2021). *Image schemas and concept invention: Cognitive, logical, and linguistic investigations.* Springer.

Howes, D. (Ed.). (1991). *The varieties of sensory experience.* University of Toronto Press.

Johnson, M. (1987). *The body in the mind: The bodily basis of meaning, imagination and reason.* University of Chicago Press.

Kant, I. (1781). *Critique of pure reason.* Internet Archive. https://archive.org/details/critiqueofpurereason_201907

Koffka, K. (1921). *The growth of the mind.* Routledge and Kegan Paul.

Koffka, K. (1935). *Principles of Gestalt psychology.* Harcourt, Brace.

Köhler, W. (1925). *The mentality of apes.* Routledge and Kegan Paul.

Lakoff, G. (1979). The contemporary theory of metaphor. In A. Ortony (Ed.), *Metaphor and thought* (pp. 202–251). Cambridge University Press.

Lakoff, G. (1987). *Women, fire and dangerous things: What categories reveal about the mind.* University of Chicago Press.

Lakoff, G., & Johnson, M. (1980). *Metaphors we live by.* Chicago University Press.

Lakoff, G., & Johnson, M. (1999). *Philosophy in the flesh: The embodied mind and its challenge to Western thought.* Basic.

Lakoff, G., & Núñez, R. (2000). *Where mathematics comes from: How the embodied mind brings mathematics into being.* Basic Books.

Langacker, R. W. (1987). *Foundations of cognitive grammar.* Stanford University Press.

Langacker, R. W. (1990). *Concept, image, and symbol: The cognitive basis of grammar.* Mouton de Gruyter.

Langacker, R. W. (1999). *Grammar and conceptualization.* Mouton de Gruyter.

Leibniz, G. W. (1703). Explication de l'arithmétique binaire. In C. I. Gerhardt (Ed.), *Die mathematische Schriften von Gottfried Wilhelm Leibniz* (Vol. VII, pp. 223–227). Ascher.

Malcolm, I. G., & Sharafian, F. (2002). Aspects of aboriginal English oral discourse: An application of cultural schema theory. *Discourse Studies, 4,* 169–181.

Mandler, J. M., & Pagán Cánovas, C. (2014). On defining image schemas. *Language and Cognition, 6,* 510–532.

Marcus, S. (2012). Mathematics between semiosis and cognition. In M. Bockarova, M. Danesi, & R. Núñez (Eds.), *Semiotic and cognitive science essays on the nature of mathematics* (pp. 98–182). Lincom Europa.

Mathematics Stack Exchange. (2015). *Euler's solution of seven bridges of Königsberg in layman terms.* https://math.stackexchange.com/q/1173328

McGreal, I. P. (2000). The paradoxes of Zeno. In J. K. Roth (Ed.), *World philosophers and their works.* Salem Press.

Montessori, M. (1912). *The Montessori method.* Frederick A. Stokes.

Murray, D. J. (1995). *Gestalt psychology and the cognitive revolution.* Prentice Hall.

Nishida, H. (1999). Cultural schema theory. In W. B. Gudykunst (Ed.), *Thinking about intercultural communication* (pp. 401–418). Sage Publications.

Núñez, R., Cooperrider, K., & Wassmann, J. (2012). Number concepts without number lines in an indigenous group of Papua New Guinea. *PLoS One, 8*(10), 1371.

Oakley, T. (2010). Image schemas. In D. Geeraerts & H. Cuyckens (Eds.), *The Oxford handbook of cognitive linguistics.* Oxford University Press.

Otten, M., Heuvel-Panhuizen, M. v. d., & Veldhuis, M. (2019). The balance model for teaching linear equations: A literature review. *International Journal of STEM Education, 6,* 30. stemeducationjournal.springeropen.com/articles

References

Paoletti, T. (2011). Leonard Euler's solution to the Konigsberg bridge problem. *Convergence.* https://old.maa.org/press/periodicals/convergence/leonard-eulers-solution-to-the-konigsberg-bridge-problem

Petersson, J., & Weldemariam, K. (2022). Prime time in preschool through teacher-guided play with rectangular numbers. *Scandinavian Journal of Educational Research, 66,* 714–728.

Piaget, J. (1952). *The origins of intelligence in children.* International Universities Press.

Piaget, J. (1969). *The child's conception of the world.* Littlefield, Adams & Co.

Popper, K. (1934). *The logic of scientific discovery.* Science Editions.

Postnikoff, D. L. L. (2014). *Metaphor and mathematics.* University of Saskatchewan Doctoral Thesis. harvest.usask.ca/server/api/core/bitstreams/c9d138b3-05a7-4965-a7b4-b142d526b2e2/

Talmy, L. (1981). *Force dynamics.* Paper presented at Conference on Language and Mental Imagery. May 1981, University of California, Berkeley.

Talmy, L. (1988). Force dynamics in language and cognition. *Cognitive Science, 12,* 49–100.

Vygotsky, L. S. (1962). *Thought and language.* MIT Press.

Wertheimer, M. (1923). Untersuchungen zur Lehre von der Gestalt, II. *Psychologische Forschungen, 4,* 301–350.

Williams, R. F. (2010). Image schemas in clock-reading: Latent errors and emerging expertise. *Journal of the Learning Sciences, 21,* 1–31.

Chapter 3
Related Processes

3.1 Introduction

Seen as individual constructs, image schemas can hardly provide a veritable explication of how mathematical ideas are generated or acquired. As isolated mental Gestalten, they would remain mere memory trace devices to specific mathematical ideas. The overall theory thus includes a whole set of interrelated processes in which image schemas play specific cognitive roles in the production of math cognition (and other forms of cognition). The main ones are: mapping, framing, layering, blending, clustering, radial structuring, and looping. These will constitute the subject matter of this chapter. Constituting a specific type of cognitive architecture, image schema theory can be seen to have implications for AI and neural network theory, as will be discussed in the next chapter (Amant et al., 2006). The term "architecture" in this case does not imply fixed structures, but adaptive ones that work via the subjective activation of the various processes by the mind of the individual mathematician.

The starting point for considering image schemas as part of a larger apparatus of interrelated processes was Lakoff's treatment of the English word *over* in his book, *Women, Fire and Dangerous Things* (Lakoff, 1987a), based on Claudia Brugman's 1981 thesis, *The Story of Over*. Lakoff showed that the many meanings and sociolinguistic uses of *over* were not autonomous but interrelated conceptually and structurally in terms of what he called "radial" category structure—which in this chapter will be adopted to indicate how image schemas interrelate with each other and with mapping processes. As Lakoff showed, the word *over* could be connected conceptually to various spatial image schemas, including a contact one, whereby the trajector could be seen to be in contact with a landmark, as in "The alpinist climbed over the mountain," or not, as in "The helicopter flew over the mountain." Lakoff also identified a group of image schemas that occur in tandem, such as the rotation and path ones, in assigning meaning to *over*, as evidenced in "Spider-man climbed all over the partition." In this chapter, this pattern will be reframed as "clustering," anticipated by Johnson (1987), who showed that a specific image schema can occur in clusters with other schemas, revolving mentally around a unique referent, such as a mathematical idea. The clusters thus form conceptual networks with referentially related nodes within it.

As a concrete initial example drawn from common everyday speech, consider the abstract concept of *ideas* as the result of mappings from source domains such as the following (Danesi, 2023a): *sight* ("I cannot see what you are saying"), *geometry* ("The views of Plato and Descartes are parallel in many ways"), *plants* ("That theory has deep roots in all philosophical traditions"), *buildings* ("Your theory is well constructed"), *food* ("That is an appetizing idea"), *fashion* ("His theory went out of style years ago"), and *commodities* ("You must package your ideas differently"), among others. These result by clustering source domains around the target domain of *ideas*—indeed, the utterances would be meaningless or bizarre to someone who does not possess the network of source domains that clusters them around a specific target domain.

3.2 Mapping

Mapping is the process that interlinks a source domain to a target domain, guided ontologically by specific image schemas or combinations of image schemas. So, for instance, the source domain of countable objects is mapped onto the target domain of arithmetic via the container image schema, to produce the conceptual metaphor of *arithmetic as a collection of numerical objects* (in containers of various sorts and sizes). This describes the conceptual structure of numbers in relation to other object collections, as in set theory—a theoretical explication that seems to fit the organization of concepts across mathematical fields.

Mapping involves the activation of specific neural circuits, as Lakoff (2014) has suggested: "The human brain is structured by thousands of embodied metaphor mapping circuits that create an extraordinary richness within the human conceptual system. They largely function unconsciously." The circuits link distinct brain regions, allowing image-schematic forms from one region to apply to another region. As Lakoff (2014) puts it:

> Motion, for example, is characterized both via topographic maps of the visual field in which activation moves across the visual map, coordinated with executing circuitry for moving the body from an initial location, through a course of motion, to a final location. The embodiment circuitry for different primitive concepts makes use of different parts of brain, which are anatomically organized by links to the body…We hypothesize that primitive concepts have a schema structure that mediates between embodiment circuitry and complex concepts.

Complex concepts are formed by a scaffolding or cascading of mapping processes, which involve image schemas in different parts of the brain. As Susac and Braeutigam (2014) concluded from their comprehensive review of the neuroscientific literature in this area, the mapping process involves "the coordinated action of

3.2 Mapping

many brain regions," which entails projecting "distinguishable functional modules onto anatomically separate brain regions." The mapping process does not, however, have to be restricted to a source-to-target domain process but can also involve a mapping of image-schematic structure directly onto representational forms, such as diagrams. So, for example, graph theory was guided initially by reifications of the meta-image-schemas of stasis, motion, and force, which enfold subtypes such as the path image schema, as discussed in Chap. 2. It can be said that the situation at hand—the actual physical configuration of the scene containing the seven Königsberg bridges layout—activated mapping circuitry based on the path schema subtype leading to the conception of a diagram representing the situation as a connected network. Image-schematic thinking in this case can be seen to have been mapped directly onto a diagrammatic source domain. It is the result of such mapping that eventually led to the conceptual metaphor of graph theory.

Sometimes, the mapping process involves already-established conceptual metaphors that are projected onto novel target domains. Consider the number line conceptual metaphor as a case-in-point. As discussed, it results from mapping the phenomenon of locomotion along a real path—moving forward and backward—onto a diagram consisting of a linear sequential layout of numbers as points or objects on an imaginary path. From this diagrammatic form, further image-schematic reasoning can be seen to unfold—right-oriented motion in space involves gaining territory, which is why positive numbers are located to the right of 0, gaining imaginary territory, while negative numbers are located to the left of 0, losing imaginary territory. The mapping between physical motion along paths onto abstract motion along a number line is hardly an act of comparison—it is the result of imagining one as equivalent conceptually to the other. Now, the question becomes: Where would we locate $\sqrt{2}$ on the number line? This is where the Pythagorean theorem comes into play as a source domain itself that is mapped (geometrically and ideationally) onto the number line. Since $\sqrt{2}$ appears as the length of the hypotenuse in an isosceles right triangle with the equal sides equal to 1, it can be located on the number line via a simple geometric construction technique, shown below (Fig. 3.1).

This exemplifies a mapping of a mapping, namely the mapping of the Pythagorean theorem, itself derived from a previous mapping, onto the number line, an already-established conceptual metaphor. This shows that there are different layers or orders of mappings, as will be discussed below. The first order involves source domains guided by image schemas derived from the experience of locations, movements,

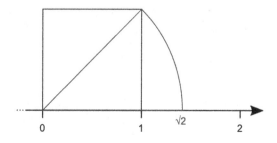

Fig. 3.1 Locating $\sqrt{2}$ on the number line

shapes, substances, containers, paths, and so on. This first-order of source domains is then mapped onto a first layer of conceptual metaphors, such as the number line, which then become themselves source domains for further mappings, allowing for the collocation of irrational numbers on the number line.

Consider, as another example, the coordinate system. As discussed, it is constructed with two number lines crossing at right angles, constituting therefore a mapping of a first-order conceptual metaphor, the number line based on the path schema, onto a second-order conceptual metaphor, the plane, guided by the verticality schema blended with the horizontality schema. The mapping process in this case produces a diagram characterized by four quadrants where points can be located. From this, we can now name the parts of the plane, starting with the *x-axis*, the *y-axis*, and the *origin*. With this new metaphorical model, we are now able to describe numbers as ordered pairs (x, y) of coordinates, representing linear motion, left-right (x) and up-down (y), from the origin (0). This model can now be used to expand mathematics; indeed without it notions such as functions, the calculus, complex numbers, and many others could never have been contemplated. As Lakoff and Núñez (2000: 38–39) maintain, this level of complex thinking was made possible ultimately by primitive, first-order image schemas, or more precisely, as argued here, meta-image-schemas, which underlie the ideational source of constructs such as the number line, which itself consists of a source (the starting point), the path itself, and the end goal (as for example a particular number)—a system of thinking applied to the coordinate or Cartesian plane:

> The Source-Path-Goal schema is ubiquitous in mathematical thought. The very notion of a directed graph, for example, is an instance of the Source-Path-Goal schema. Functions in the Cartesian plane are often conceptualized in terms of motion along a path—as when a function is described as "going up," "reaching" a maximum, and "going down" again.

As Lakoff and Núñez (2000: 43–44) go on to elaborate, primitive image schemas do not occur as isolated notions, but rather, as guiding mappings in specific ways. So, a further mapping based on the meta-image-schema of motion, blended with the path and verticality derived schemas, can be seen to produce the complex plane, in which operations on complex numbers are visualized as motions through the Cartesian space, with the *y-axis* envisaged as a vertical Imaginary number line (Im) and the *x-axis* as a horizontal Real number line (Re) (Fig. 3.2).

Fig. 3.2 The complex number plane

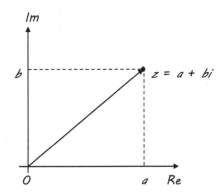

The result is that we can now locate a complex number, such as $z = a + bi$, with a the real part and bi the imaginary part in the plane, as well as its vectorial (directional) features in terms of the angle between the hypotenuse line above and the Real axis. Besides the fact that complex numbers have many practical applications, especially in physics and engineering, in "mathematics" they expanded the notion of number considerably, allowing for a view of real numbers as a subcategory of complex numbers. Moreover, it is now known that the set of complex numbers constitutes a *field*, which means, essentially, that: (a) any two complex numbers can be added and multiplied to produce another complex number; (b) for any complex number, z, its additive inverse, $-z$, is also a complex number; and (c) every nonzero complex number has a reciprocal complex number. The study of the functions of complex variables has led to another branch of mathematics, *complex analysis*, which has had many uses and applications within "mathematics" and outside of it in scientific fields. In effect, without the brain's mapping circuitry, which unfolds at higher and higher levels of complexity, such notions, branches, and theories would not have crystallized.

Lakoff (2014) called the layering of different orders of mappings a "cascading" system, which has also been called "scaffolding" (Veale & Keane, 1992). As Lakoff puts it:

> A theory of cascades is necessary for two reasons: In complex concepts that make use of multiple primary concepts and primary metaphors, there will be a multiplicity of embodiment. Cascade theory provides the circuitry necessary to carry this out. it also provides the circuitry necessary to link the embodiment of linguistic form (in sound, writing, sign, and gesture) to the embodiment of meaning.

3.3 Framing

Early work by linguist Michael Reddy (1979) on conceptual metaphors indicated that some are rudimentary or first-order ones, based on primitive image schemas or meta-schemas as they have been called here. From Reddy's analysis, the notion of *metaphorical framing* crystallized, which implies the use of established conceptual metaphors to frame further conceptualizations via cascading mappings and the appurtenant blending of image schemas. Extending this notion into the realm of arithmetic, it can be seen, for instance, that the number line conceptual metaphor becomes itself a "metaphorical frame" for envisioning arithmetical ideas and symbols—for example, a positive sign (+) designates "go right," while a negative one (−) "go left." Now, arithmetical operations can be understood conceptually by using the number line as a frame system. In image-schematic terms, it is obvious that the path–source–goal schema is involved along with a horizontality schema (left-versus-right). For example, adding (+2) and (+3) means go right two units from zero (the start point) and then three more to the right from there. The end point (goal) is (+5) (Fig. 3.3).

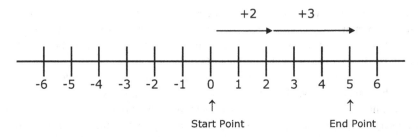

Fig. 3.3 Adding positive numbers on the number line as a frame

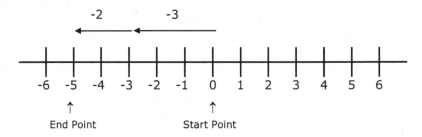

Fig. 3.4 Adding negative numbers on the number line frame

Needless to say, because of the law of commutation, the order of the movements is not relevant—one could go 3 units to the right first and then 2 units more to reach the end point. Also, note a general property of framing based on the number line. The three meta-image schemas are inherent in all framing procedures—stasis is seen in the fact that specific points on the line are reached, motion in the fact that there is movement from point to point, and force in the fact that the movement is directed in specific ways. Since this is characteristic of all frames based on the number line, it will not be repeated in the subsequent descriptions.

Adding two negative numbers is framed as motion in the opposite direction. For example, (−3) + (−2) tells us to go left three units from 0 and then left two more from there. The end point is (−5) (Fig. 3.4).

Again, the motion could have occurred by commuting the numbers. Adding numbers with different signs is framed as a blending of motion and rotation schemas. Consider (+5) + (−7). The first number tells us to go right five units from 0. At that point, a rotation occurs whereby the second number tells us to change direction and go left seven units from there. The end point is (−2) (Fig. 3.5).

Notice that the sign of the number with the larger absolute value is also the sign of the answer (end point). Multiplication is framed in terms of the iteration image schema, whereby the same value of a specified number is repeated. For instance, (+2) × (+3) means that 2 units are repeated three times to the right of 0—the same result (end point) would be obtained with (3) × (2), because of the law of commutation (Fig. 3.6).

3.3 Framing

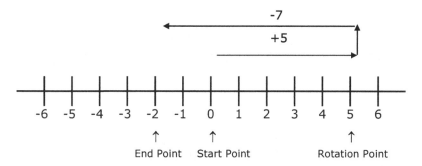

Fig. 3.5 Adding differently signed numbers on the number line frame

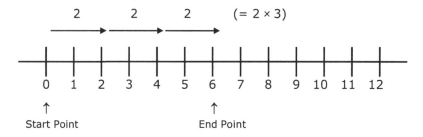

Fig. 3.6 Multiplying numbers via the number line frame

Now, consider multiplying negative numbers with the number line frame. If the signs of the two are different, then their product is negative. For example, $(-3) \times (+2) = (-6)$, because this states that (-3) must go in the "same direction" (to the left) twice, constituting the result of the iteration image schema (Fig. 3.7).

Division is based on a blend of part-whole, iteration, and equilibrium image schemas, since it involves splitting the operation into equal units as the numbers move along the number line. So, the division of positive integers can be framed as equal units that move iteratively toward the 0 point from the right. For example, $(6) \div (3)$ indicates the leftward movement of 3 equal units on the line from 6 to 0. As can be seen, each is 2 units long, which is the answer (Fig. 3.8).

These frames inform widely used pedagogical techniques, pointing anecdotally to their psychological validity. As Jeff Frykholm (2010) has shown insightfully, the number line, as the title of his book puts it, *Learning to Think Mathematically with the Number Line*, is an effective learning frame because it taps into the primitive image-schematic basis of arithmetic, allowing learners literally to develop "ways of seeing" arithmetical procedures. As he notes (Frykholm, 2010: 7): "The number line allows students to engage more consistently in the problem as they jump along the number line in ways that resonate with their intuitions. While they are jumping on the number line, they are able to better keep track of the steps they are taking, leading to a decrease in the memory load otherwise necessary to solve the problem."

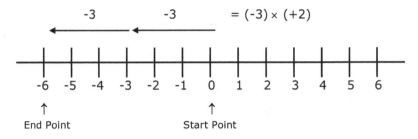

Fig. 3.7 Multiplying negative numbers via the number line frame

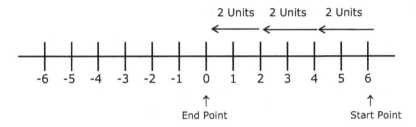

Fig. 3.8 Division on the number line as a frame

This does not mean that all students will understand the number line as a frame for grasping arithmetical operations. It is an abstract model that will be beyond comprehension at the "pre-math" stage. It can only be conceptualized at "math" and "mathematics" levels. In a relevant study, Tsang et al. (2015) found that the mirror symmetry that is evident between negative and positive integers around the zero point on a number line may not be easily understandable even to older children, at first. The researchers focused on getting the students to notice the symmetry, so as to be able to impart to them that movement along the line in different directions from zero produced the different categories of number. At first, the students indicated difficulties in understanding. But, with appropriate prompts, they eventually came to grasp the relation between the numbers as laid out symmetrically along the line, suggesting that at a "math" stage of learning, appropriate pedagogy will easily get students to grasp the meaning and uses of this complex conceptual metaphor.

Overall, the work in math education has shown that number lines assume a vast variety of forms and pedagogical functions, depending on the age of the students. They include the following: linear arrangements of countable objects; number paths, such as strips laid out on the floor, with countable spaces; unmarked number lines requiring students to subdivide them appropriately with numerals; double number lines that relate to the learning of ratio and proportion; among others (Bobis, 2009; Drake, 2014). Some educators use vertical number lines in tandem with horizontal ones to support the conceptualization of graphing in early elementary school and of coordinate grids in later grades.

3.4 Layering

The process of subsequent mappings—mappings of mappings—can be called layering (Danesi, 2023a, 2023b). The process produces complex concepts via a chronologically based interrelated system of mappings of source domains. This can be explained as follows:

First-Order Layer
A first-order layer of conceptual metaphors, such as the number line, results from the mapping of a primitive source domain that is formed directly by an image-schematic thought pattern derived from the physical experience of movement along a path onto an arithmetical target domain. Many first-order concepts are produced in this way, such as the notion of a circle as a geometrical object—a source domain which results from a mapping of a primitive source domain based on the observation of circularity in the world (the sun, the moon, for example) onto the target domain of geometry, where it can then be examined for inherent properties.

Second-Order Layer
The target domains produced by first-order mapping processes then become themselves source domains that can mapped onto new target domains, producing second-order conceptual metaphors such as the coordinate system. Significantly, these retain first-order image-schematic structure, which is blended with other schematic configurations, such as, for example, the horizontality and verticality schemas, which guide the mapping pattern to produce the coordinate plane. Many concepts in mathematics are second-order ones. For instance, the first-order source domains constituting basic geometrical concepts, such as the spatial image-schematic relationships between the angles and sides of triangles, when mapped, produce the target domain of trigonometry as a new conceptual system.

Third-Order Layer
The second-order layer of mathematical concepts, once formed, becomes itself a source domain that is mapped to produced increasingly complex third-order mathematics. For example, a mapping of the coordinate system produces analytic geometry, the calculus, functions, and other related concepts. Again, these new orders retain the image-schematic structure of previous orders, combining them in various new ways.

To summarize, primitive image schemas, such as the path and container ones, which are themselves guided by meta-imagistic cognition (Chap. 2), inform the constitution of the initial source domains that are mapped onto target domains that make up the first order of concepts. These then become source domains themselves that are mapped onto target domains that expand the concepts into more complex second-order ones. These, in turn, provide new source domains which, when mapped further, result in a third-order layer of mathematical knowledge, such as complex number theory. Throughout the layering process, image-schematic structure is preserved albeit in blended ways (discussed below, Turner, 2012).

As an example of how layering processes might unfold historically, consider the zero concept (Aczel, 2015; Kaplan, 2000; Seife, 2000). The zero emerges as a place holder in various ancient numeral systems, standing for "empty" (without value). This suggests that the path schema was operative in establishing this meaning at a primitive cognitive level. In the Babylonian and Sumerian systems, for instance, a blank (empty) space was used for this concept in numeral layouts (that is, in path-like configurations). As a first-order concept, it can be thus surmised that the zero notion is derived from the experience of observing empty spaces in the physical layouts of objects, as was the case, for instance, in ancient astrological language which labeled the spaces between bodies in the sky as "empty." The word *zero*, in fact, derives from *ziphirum*, a Latinized form of the Arabic word *sifr* which, in turn, is a translation of the Hindu word *sunya* (void or empty). So, at this initial layer of mathematical meaning, the zero concept derives from a mapping of the sense of empty space onto arithmetic. From this, an actual symbol for zero emerges in various numeral systems, including the Mayan, Indian, and Chinese ones around the fourth and fifth centuries CE—"filling," the empty space with a mathematical object. This filling practice led to a second-order conceptualization of the zero concept as an actual number that could be used within calculations—a mapping of a previous mapping. This stage is traced generally to the Indian mathematician Aryabhata in the fifth century CE, a conceptualization that was established further into mathematics by another Indian mathematician, Brahmagupta. From his work, the notion of negative numbers then crystallized. For the sake of historical accuracy, it should be noted that the negative number concept surfaced as well in ancient China in *The Nine Chapters of the Mathematical Art* (Yong, 1994), but it was not treated, as best as can be told, as a concept related to zero, before Brahmagupta (Kaplan, 2000).

The circle symbol for zero was finally embedded into mathematics in 820 CE, when the Persian mathematician Al-Khwarizmi, in his book, *On the Calculation with Hindu Numerals*, showed how it could be used in various ways within arithmetic (Daffa, 1977). This second-order layer of conceptualization, along with the notion of negative numbers, did not reach most of Europe until the early thirteenth century, when Leonardo Fibonacci from the Republic of Pisa published his *Liber Abaci* in 1202. Fibonacci succeeded in convincing his readers that the decimal system, which he had learned about during his visits to the Middle East, was far superior to the Roman one (Devlin, 2011). But Fibonacci realized that a symbol for "nothingness" would bring about philosophical objections. So he started off his book reassuring readers that zero was only an arithmetical sign that allowed for all numbers to be written (cited in Posamentier & Lehmann, 2007: 11):

> The nine Indian figures are: 9 8 7 6 5 4 3 2 1. With these nine figures, and with the sign 0, which the Arabs call zephyr, any number whatsoever is written.

It was at this point in time that the third-order layer was forming in mathematics broadly, as the zero concept was being mapped onto various new target domains, leading to a new set of axioms, definitions, and theorems for arithmetic, culminating much later in Peano's (1908) axioms, which start with defining 0 as a natural number. In effect, such uses of the zero concept became possible after becoming established as a third-order concept, with its properties identified and classified at this level. Below is a partial list (Table 3.1).

Such properties would not have been thinkable without the layering process, which gradually produced system thinking within a particular area of mathematics. New layers of mathematical reasoning are built on lower-order layers, even though this might not appear evident at first consideration. Note that it is impossible to discuss

3.4 Layering

Table 3.1 Properties of zero

Operation	Property	Examples
Addition	$x + 0 = x$	$5 + 0 = 5$
Subtraction	$x - 0 = x$	$5 - 0 = 5$
Multiplication	$x \times 0 = 0$	$5 \times 0 = 0$
Division	$x \div 0 =$ undefined	–
Exponentiation	$x^0 = 1$	$5^0 = 1$
Root	$\sqrt{0} = 0$	$\sqrt{0} = 0$
Logarithm	$\log(0) =$ undefined	–
Factorial	$0! = 1$	$0! = 1$
Sine	$\sin 0^0 = 0$	$\sin 0^0 = 0$
Cosine	$\cos 0^0 = 1$	$\cos 0^0 = 1$
Tangent	$\tan 0^0 = 0$	$\tan 0^0 = 0$
Derivative	At $x = 0$, the derivative is 0	At $x = 0$, $dy/dx = 0$
Integral	$\int 0 \, dx = c$ (the constant)	$\int 0 \, dx = c$

mathematics, or any other abstract system, without metaphor—hence terms such as "mapping" and "layering." This indirectly corroborates Lakoff and Johnson's (1980) view that there is no discourse about abstractions without recourse to metaphor.

The layering process as such resonates intuitively with teachers of elementary school mathematics. For example, Maas (2023) explored the notion of zero in young children as a process of assigning mathematical meaning to something they expected not to be there. The research found that the emptiness meaning was intuitive in children, indirectly corroborating its presence as a first-order concept:

> Do children know that nothing will be left after everything is taken away? Our studies showed that 12-month-olds could express that something was missing by pointing to where it should have been, and 4-year-olds were able to search for objects that had gone missing. Preschoolers understood quantities up to nine, including zero, or the empty set. When asked to label containers with five cookies, half a cookie, and no cookies, children accurately placed a blank sign on the "No Cookies" jar to show that there were zero cookies in that container. They still understood the blank sign's meaning when asked again two weeks later.

This understanding coincides with the empty container schema as a notion of something missing that would typically be there. When problems of understanding emerge in elementary children with regard to the zero symbol (0), it can be attributed to the fact that it is presented as a second-order concept that requires some form of concrete pedagogical intervention—that is, children struggle at first to understand that it is a number with a position on the number line. Significantly, as Maas (2023) found, language plays an obstacle role in the learning process: "Although preschoolers are capable of understanding empty sets, some struggle with the term *zero*." In one specific experiment, Maas found that children became confused when asked to give "zero balls to a bird." However, when the same task was phrased as, "Do not give any balls to the bird," the children easily understood the directions. Through relevant pedagogy, the study concludes, children eventually developed their understanding of zero.

To reiterate, what makes complex mathematical concepts understandable, if learned sequentially—first-order ones learned before second-order ones and then before third-order ones—is that subsequent mappings preserve image-schematic structure, revealing the principle of invariance (Lakoff, 1979). The higher the density of layering, the more intrinsic are the new target domains to "mathematics," the discipline. As Lakoff and Núñez (2000: 44) put it:

> What will eventually become an abstract concept is formed as a metaphor; further concepts are derived from the primary metaphors and by other tropical processes. The first abstract concepts were formed in an identical way—it is in origination concepts that the same kinds of image schemas are found archeologically, such as the *path* schema. Each metaphorical layer carries inferential structure systematically from source domains to target domains—systematic structure that gets lost in the layers unless they are revealed by detailed metaphorical analysis.

As discussed in the previous chapter, the hypothesis put forward here is that image-schematic processes are ultimately guided by the meta-image-schemas of *stasis, motion*, and *force*, which can be seen to be operative in various ways in layering processes. For example, the *stasis* schema underlies the constitution of source domains that lead to conceptual metaphors such as *states are locations, states are bounded regions*, which involves thinking of numbers as having a position in bounded spaces (Lakoff & Núñez, 2000). This meta-image-schema is operative in the specific formation of the symmetry schema, which is itself operative in the conceptualization of zero as a position on the number line that divides positive and negative numbers symmetrically—to the right and the left, respectively. The *motion* meta-image-schema underlies mapping processes that produce conceptual metaphors such as *changes are movements, moving objects in and out of bounded regions*, and others. Even the way we talk about numbers in terms of motion reveals its unconscious operationality in guiding thought—this is why we say "move right three units along the *x-axis*," "go up five units along the *y-axis*," and so on. The *force* meta-image-schema describes some end result of conceptual change initiated by the motion schema; it is the basis of branches such as the calculus. In ordinary descriptive language, it manifests itself in utterances based on the verb "*must*," such as "You *must* change the sign in order to change directions on the number line" or "To reach the end of the series, you *must* figure out the underlying structure of the series first."

Interestingly, the metaphor of "forcing" was introduced into set theory as a technique for proving consistency of results (Cohen, 1966), and thus as a means to expand the Zermelo–Fraenkel continuum hypothesis, which introduced the notion of "pure sets," as sets preventing the inclusion of urelements—that is, of sets containing elements that are not themselves sets, so as to avoid circularity. The force schema in this case involves conceptually "forcing" elements out of a mathematical container, which in more precise mathematical language means that certain statements do not belong in certain sets.

3.5 Blending

As Lakoff argued at his Fields lecture (Chap. 1), Gödel's (1931) famous proofs were inspired by Cantor's diagonal proof method (Danesi, 2011). In simplified terms, Gödel proved that within any formal logical system there is a statement, S, in the set

3.5 Blending

of statements within the system, that could be extracted from it by going through them in a diagonal fashion—called Gödel's diagonal lemma—that can be neither proved nor disproved.

As Lakoff claimed, the Gödelian proof is a classic example of how ideas that may seem independent are actually interconnected, an awareness which occurs when the brain identifies two distinct entities in different neural regions as the same entity in a third neural region—called a "blending" process. This process was activated by the diagonal and zig–zag configuration of the path image schema inherent in Cantor's original proof. When Gödel adopted this schema as part of his own proof, the blending process produced the relevant metaphorical insight: Can the same type of proof be mapped on to the problem at hand? As it turned out, the answer was a positive one. As Lakoff and Núñez (2000) observe, examples such as this one show concretely that "understanding mathematics requires the mastering of extensive networks of metaphorical blends." Blending overlaps with processes such as mapping and layering.

One of the first mentions of this notion is found in Mark Turner's book, *The Literary Mind* (1997: 93), in which he states that "Conceptual blending is a fundamental instrument of the everyday mind, used in our basic construal of all our realities, from the social to the scientific." Blending theory was elaborated formally a little later by Fauconnier and Turner (1998, 2002, 2003), providing a plausible explanatory framework of how inputs from different regions in the brain, based on a common experiential image-schematic basis (for example, the container schema), are amalgamated to form a new concept, which is much greater than the sum of its inputs. From this, new properties emerge and new structural–conceptual relationships are created that did not exist in the original inputs. Below is the Fauconnier–Turner model of conceptual blending (Fig. 3.9).

In the generation of the number line conceptual metaphor, for example, the input space 1 is the source domain formed on the basis of the path schema and input space

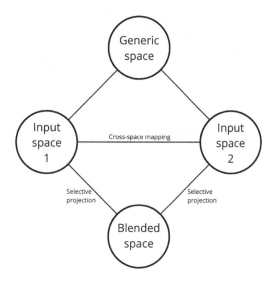

Fig. 3.9 Conceptual blending (Fauconnier & Turner, 2002)

2 is the source domain based on the image schema of numbers as objects in a container. There is then a cross-mapping which produces the new blended space (the number line as a new concept). The selective projection feature can be interpreted as the intervention of the subject (the mathematician, for example) who is inferring the cross-space mapping and the generic space as the neuro-cognitive frame in which the whole process unfolds, providing the input spaces. This makes explicit what happens in the brain as layers of metaphorical concepts are formed on the basis of image-schematic connectivities and related mappings. To cite Lakoff and Núñez (2000: 48):

> A conceptual blend is the conceptual combination of two distinct cognitive structures with fixed correspondences between them...When the fixed correspondences in a conceptual blend are given by a metaphor, we call it a metaphorical blend. An example is the Number-Line Blend, which uses the correspondences established by the metaphor *Numbers Are Points on a Line*. In the blend, new entities are created—namely, number-points, entities that are at once numbers and points on a line.

It has been found, significantly, that images, both novel and recalled, are processed by the visual cortex in the occipital lobe, spanning both cerebral hemispheres. This kind of neuroscientific finding suggests that blending, based on image-schematic cognition, unfolds in regions where visual imagery and perception are processed (Brown, 2018; Kutter et al., 2022). The question becomes: How is the blending process activated? As Turner suggests, the inspiration comes from the "creative inferences" of the individual "blender" (Turner, 2012, 2014), which recalls what Charles Peirce called "abductions" or individual acts of insight (1931–1958, vol. 5: 180):

> The abductive suggestion comes to us like a flash. It is an act of *insight*, although of extremely fallible insight. It is true that the different elements of the hypothesis were in our minds before; but it is the idea of putting together what we had never before dreamed of putting together which flashes the new suggestion before our contemplation.

Alexander (2012: 28) aptly uses the notion of creative inference inducing neural blends to describe the emergence of negative numbers. Note that his use of the term "collapse" is akin to the notion of layer and that his use of the pronoun "we" is an indirect reference to the creative mind of the mathematician:

> Using the natural numbers, we made a much bigger set, way too big in fact. So we judiciously collapsed the bigger set down. In this way, we collapse down to our original set of natural numbers, but we also picked up a whole new set of numbers, which we call the negative numbers, along with arithmetic operations, addition, multiplication, subtraction. And there is our payoff. With negative numbers, subtraction is always possible. This is but one example, but in it we can see a larger, and quite important, issue of cognition. The larger set of numbers, positive and negative, is a cognitive blend in mathematics...The numbers, now enlarged to include negative numbers, become an entity with its own identity. The collapse in notation reflects this. One quickly abandons the (minuend, subtrahend) formulation, so that rather than (6, 8) one uses −2. This is an essential feature of a cognitive blend; something new has emerged.

In effect, new concepts emerge through "collapsing" processes in the brain of the mathematician. Conceptual metaphors that surface at the highest, or third-order,

3.5 Blending

layer of derivation (mapping) are the result of blends built upon blends. Recall Euler's discovery of graph theory. Within the Fauconnier–Turner framework, it can be envisioned as a blending of the stasis, motion, and force meta-image-schemas, which constitute the initial generic space (a holding pattern of imagistic thoughts) that undergirds inputs such as paths and connecting points between paths in a network which when blended together via Euler's creative inferences, produce the concept of graphs as mathematical systems (Chap. 2). René Thom (1975, 2010) called this kind of discovery a "catastrophe," because it overturns existing knowledge, emerging by happenstance through the blending of thought forms.

One never knows when and to whom the creative inference (or abduction) will come that will induce a blend. Consider a well-known anecdote that French mathematician Henri Poincaré recounted in his book, *Science and Hypothesis* (1902). Poincaré had been puzzling over a problem that seemed to defy a solution, leaving it aside to embark on a geological expedition. As he was about to get onto a bus, the crucial idea for a solution came to him in a flash of insight (an abduction). He claimed, afterward, that without this flash, the solution would have remained unknown to him. As Poincaré elaborates, he became aware that the solution inhered in blending Fuchsian functions with non-Euclidean geometry (Poincaré, 1902: 23):

> Just at this time I left Caen, where I then lived, to take part in a geologic excursion organized by the École des Mines. The circumstances of the journey made me forget my mathematical work; arrived at Coutances we boarded an omnibus for I don't know what journey. At the moment when I put my foot on the step the idea came to me, without anything in my previous thoughts having prepared me for it; that the transformations I had made use of to define the Fuchsian functions were identical with those of non-Euclidean geometry. I did not verify this, I did not have time for it, since scarcely had I sat down in the bus than I resumed the conversation already begun, but I was entirely certain at once. On returning to Caen I verified the result at leisure to salve my conscience.

Research on blending theory has produced some intriguing findings that provide corroborative evidence as to the psychological validity of image schema theory and conceptual metaphor mappings. For example, Guhe et al. (2011) developed a computational model which had the ability to come up with different conceptualizations of number by blending together common features. The model was based on Lakoff and Núñez's conceptual metaphors for arithmetic, focusing on the container image schema as the conceptual link between number-based target domains. The algorithm searched for commonalities (inputs) and then transferred (mapped) them from one domain to another, producing mathematical conceptual blends. As Fauconnier and Turner (2002) cogently argued and illustrated, blending creates continuous intelligence—which inheres in a blending upon blending process, prompted by creative inferences. Reading a math theorem in a book might lead some individual mathematician to devise another one or to use it as part of some new idea, based on the individual's experiences and background knowledge related to the theorem. When others adopt it in order to develop it further, the idea becomes a shared one and thus part of "mathematics" as a discipline. The modus operandi of mathematicians is, seemingly, to build upon ideas created through blending processes. In this way, they construct entire "stories," smaller and larger, that make up mathematics as a whole. To quote Turner (2005):

> As long as mathematical conceptions are based in small stories at human scale, that is, fitting the kinds of scenes for which human cognition is evolved, mathematics can seem straightforward, even natural. The same is true of physics. If mathematics and physics stayed within these familiar story worlds, they might as disciplines have the cultural status of something like carpentry: very complicated and clever, and useful, too, but fitting human understanding. The problem comes when mathematical work runs up against structures that do not fit our basic stories. In that case, the way we think begins to fail to grasp the mathematical structures. The mathematician is someone who is trained to use conceptual blending to achieve new blends that bring what is not at human scale, not natural for human stories, back into human scale, so it can be grasped.

Even a simple arithmetical concept such as "7 is larger than 4" involves blending the image schema of containers of various sizes with the measurement schema, which undergird the conceptual metaphor, *numbers are collections of objects of differing sizes*. One could go through virtually all mathematical ideas with a similar explanatory method.

Blending theory has various precursors, including interaction theory, as developed by I. A. Richards (1936) and Max Black (1962), who conceived of metaphorical reasoning as a process of establishing conceptual links between what is known (source domains) and what needs to be known (target domains) on the basis of an experiential interaction between them initially (image-schematic thought). Soskice (1985) suggests, colorfully, that the two domains "animate" each other. Consider the concept of the number line once again as a mapping of the source domain informed by the path image schema, itself originating from the experience of walking, as Wallis remarked (Chap. 1). Once incorporated into the discipline of "mathematics," it became a diagrammatic model of how we count and organize counting in a sequence from small to large, such as laying out objects to show how many of them there may be with respect to other layouts (cardinality). It both mirrors and then subsequently structures the thoughts and actions we perform when we count. In effect, the number line became a source domain for further mathematics, after Wallis, leading to more complex mathematics via higher-order layering processes. These then instill increasingly abstract mathematical thoughts in mathematicians, which become part of larger and larger mathematical "stories," to use Turner's term. Without the number line, it is unlikely that they would have been "told" in the first place.

The initial layer of arithmetical thinking derives from the experience of counting, which emerges during the "pre-math" phase, and has counterparts in other species. Perceiving the "counted objects" as occurring in containers of different sizes allows us subsequently to turn intuitive counting into symbolic behavior (numerals and other symbols) that can be grasped and manipulated intellectually, as evidenced during the secondary "math" phase of mathematical learning. This pattern of abstract thinking can be called "numerality." Analogously, the initial layer of geometrical thinking derives from the experience of estimating distances and of recognizing different shapes as meaningful, called "spatiality." When it is subsequently understood that spatiality and numerality can be blended, a new sense of "mathematics" crystallizes. From this point onward, subsequent layers of understanding produce highly sophisticated mathematical abstractions in the human mind that are unique among species.

The emergence of such abstract thinking processes in various domains, not only mathematics, has endowed humans with the capacity to carry the world around in their heads, so to speak, and to transform intuitive number and word sense into reflective ones. As the historian of science Jacob Bronowski (1977: 24) insightfully put it, it is this transformation that allows humans uniquely to be able to project the future:

> The images play out for us events which are not present to our senses, and thereby guard the past and create the future—a future that does not yet exist, and may never come to exist in that form. By contrast, the lack of symbolic ideas, or their rudimentary poverty, cuts off an animal from the past and the future alike, and imprisons it in the present. Of all the distinctions between man and animal, the characteristic gift which makes us human is the power to work with symbolic images.

3.6 Clustering and Radial Networks

An important aspect of image-schematic structure is that it is not an isolated one, whereby one schema would be processed differentially from other schemas. Rather, as Lakoff (1987a) cogently argued, specific image schemas are mapped constantly onto various target domains, leading to the formation of interlinked clusters of conceptual metaphors in discourse and other cognitive systems (Cameron, 2010). This notion, which Lakoff relates to discourse and linguistic grammar, can be extended to math cognition.

For example, the notion of *number* clusters around the following image schemas, which are then mapped onto different target domains, showing how they are interconnected: *objects* (distinct quantitative units), *containers* (elements in containers), *paths* (elements on the number line), *points* (in reference to the points on a line or coordinate plane), *geometrical forms* (as in triangular numbers, square numbers, etc.), *placement* (as in decimal and binary numerals), *rotation* (as in sexagesimal systems), *orientation* (as in negative numbers represented as objects to the left of zero on a number line), *spatiality* (the location of numbers in specific conceptual spaces), etc. The clustering of these image schemas around the target domain of number produces an interlinked cognitive network, which portrays how the various meanings and functions of number in mathematics are intertwined conceptually. A partial clustering network is the following one (Danesi, 2023a, 2023b) (Fig. 3.10).

Not all networks manifest a clustering structure. Another type involves different target domains connected by the same image schema. This type of network can be called *radial,* following Lakoff (1987a, 1987b), since it can be envisioned as a single schema "radiating outward" to deliver different concepts. For example, the *path* image schema not only allows us to conceptualize number but also such other

Fig. 3.10 Clustering network for number

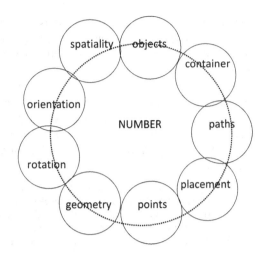

Fig. 3.11 Radial network for the path schema

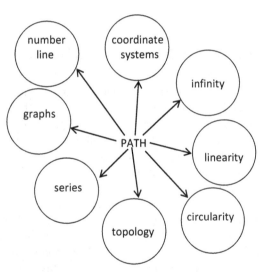

concepts as *coordinates* (intersecting number paths), *infinity* (paths that never end), *linearity* (paths that are straight rather than curved), *circularity* (circular paths), *topology* (paths that can take any shape), *graphs* (which show how certain paths are constrained by their vertex-edge structure, as in the Königsberg Bridges Problem), *number line* (which shows a linear path as consisting of equally spaced number points), *series* (which are mathematical objects laid out in some linear sequence), among many others. A partial radial network for the path schema can be shown as follows (Fig. 3.11).

Radial structure can be defined as the tendency to envisage abstract concepts as implicating each other through a specific frame of reference (a single image schema). Relevant research indicates that some mathematical conceptual metaphors are based on radial structuring, while others are more diverse, involving clustering networks (Danesi, 2023a, 2023b). Network structure in AI systems reveals the presence of conceptual interconnectivity resulting from blending, layering, and mapping processes at different levels and orders (Chap. 4). Each node (domain) in the network is an *n*-order mapping (first-order, second-order, or third-order).

3.7 Looping

One category of image schemas that is hardly ever discussed in the relevant literature, albeit alluded to indirectly, is the *looping* image schema. This is a subtype of path schema that turns on itself, as if in a loop, producing circularity of reasoning, paradoxes, and various other systems from which there is no "escape" cognitively (Hofstadter, 1979).

Looping is saliently evident in logical conundrums such as the halting problem (Turing, 1936), which is related to the notion of a "Turing machine"—a mathematical model of a hypothetical computing machine which can use a set of predefined rules to determine a result from a set of input variables. The question it begs is the following one: Can an algorithm capable of deciding whether a given statement is provable from the axioms using the rules of first-order logic be devised? To put it another way: Given a computer program and an input, will the program finish running or will go into a loop and run forever? Turing himself argued that no algorithm for solving this problem can exist logically. Here is a paraphrase of his proof by contradiction:

> Assume that there is such a program. If so, we could run it on a version of itself, which would halt if it determines that the other program never stops, and runs an infinite loop if it determines that the other program stops. This is a contradiction.

The problem was actually prefigured by David Hilbert (1902, 1926, 1928), who believed that mathematicians should never stop trying to find solutions to intractable problems in mathematics, since the effort alone is what propels mathematics. As he put it (Hilbert, 1902: 438):

> Is the axiom of the solvability of every problem a peculiar characteristic of mathematical thought alone, or is it possibly a general law inherent in the nature of the mind, that all questions which it asks must be answerable?...This conviction of the solvability of every mathematical problem is a powerful incentive to the worker. We hear within us the perpetual call: There is the problem. Seek its solution. You can find it by pure reason, for in mathematics there is no ignoramibus.

What may block the finding of solutions is the presence of looping structure in mathematical thinking. The halting problem indicates that the solution path (cognitively) turns on itself—a rotating singular path image schema. There is no escape from the loop and hence its unsolvability.

As one other example of looping, consider the *Collatz conjecture,* formulated in 1935 by German mathematician Lothar Collatz, who found that we always end up with the number one if we apply the following rule: If a number, n, is even, make it half, or $n/2$; if it is odd, triple it and add one, or $3n + 1$. If we keep repeating this rule, we always end up with the number one—hence a loop. Here is a concrete example:

Example: 12

First Application of Rule: 12 is even, so we divide it by 2
$12/2 = 6$

Second Application: 6 is even, so we divide it by 2
$6/2 = 3$

Third Application: 3 is odd, so we triple it and add 1
$(3)(3) + 1 = 10$

Fourth Application: 10 is even, so we divide it by 2
$10/2 = 5$

Fifth Application: 5 is odd, so we triple it and add 1
$(3)(5) + 1 = 16$

Sixth Application: 16 is even, so we divide it by 2
$16/2 = 8$

Seventh Application: 8 is even, so we divide it by 2
$8/2 = 4$

Eighth Application: 4 is even, so we divide it by 2
$4/2 = 2$

Ninth Application: 2 is even, so we divide it by 2
$2/2 = 1$

Now, since 1 is an odd number, then using $3n + 1$, we get 4, which is the application just before; making it half equals 2, which when divided by 2 again brings us to 1, and this process continues forever. The loop at the end can be shown with a directed graph (Fig. 3.12).

The question becomes: Is this really always the case? Is there a number where oneness is not achieved? No one has ever demonstrated that it is always true, and the reason may well be that its looping structure blocks us from devising a proof. Now, the conjecture, significantly, involves the path image schema, given that it can be modeled as a directed graph, consisting of a set of vertices connected by directed edges called *orbits*, with the final orbits producing the loop. In 1976, Riho Terras demonstrated that, after repeated application of the Collatz rule, almost all numbers eventually wound up lower than where they started on a directed graph. In 2019, Terence Tao proved that for almost all numbers the Collatz sequence of a number, n, ends up below n. In other words, for almost every number, we can guarantee that its Collatz graph goes as low as we desire it to go and this is "about as close as one can get to the Collatz Conjecture without actually solving it" (Tao, 2019).

Fig. 3.12 Looping structure in the Collatz conjecture

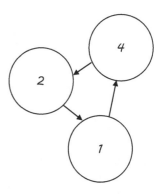

The notion of looping was examined in detail by Douglas Hofstadter (1979, 2007), who argued cogently that "strange loops" in human consciousness crystallize as the complexity of symbols in the brain accumulates to a high level. This leads to phenomena whereby, in moving through a system, such as some branch of mathematics, one finds oneself back to where one started. Looping is the source of paradoxes evident in statements such as, "This statement is false," which leads to a looping pattern of thought, caused by self-reference. Looping occurs whenever movement is made, according to Hofstadter, upward or downward through the levels of some system, unexpectedly arriving back to the starting point. Hofstadter sees this as a force (a meta-image-schema as envisioned here) which produces paradoxes in logic (such as the liar paradox), designating a system in which a strange loop appears within a "tangled hierarchy."

References

Aczel, A. D. (2015). *Finding zero*. Palgrave Macmillan.
Alexander, J. (2012). On the cognitive and semiotic structure of mathematics. In M. Bockarova, M. Danesi, & R. Núñez (Eds.), *Semiotic and cognitive science essays on the nature of mathematics* (pp. 1–34). Lincom Europa.
Amant, R. S., Morrison, C. T., Chang, Y-H., Mu, W., Cohen, R., & Beal, C. R. (2006). *An image schema language*. https://apps.dtic.mil/sti/pdfs/ADA458943.pdf
Black, M. (1962). *Models and metaphors*. Cornell University Press.
Bobis, J. (2009). The empty number line: A useful tool or just another procedure? *Teaching Children Mathematics, 13*, 410–413.
Bronowski, J. (1977). *A sense of the future*. MIT Press.
Brown, R. D. (2018). *Neuroscience of mathematical cognitive development: From infancy through emerging adulthood*. Springer.
Brugman, C. (1981). *The story of over*. University of California.
Cameron, L. (2010). The affective discourse dynamics of metaphor clustering. *Journal of Pragmatics, 42*, 97–115.
Cohen, P. J. (1966). *Set theory and the continuum hypothesis*. Dover.
Collatz, L. (1935). *Das Differenzenverfahren mit höherer Approximation für lineare Differentialgleichungen*. Thesis, University of Berlin.

Daffa, A. A. (1977). *The Muslim contribution to mathematics*. Croom Helm.
Danesi, M. (2011). George Lakoff on the cognitive and neural foundations of mathematics. *Fields Notes, 11*(3), 14–20.
Danesi, M. (2023a). *Introduction to semantics*. Lincom Europa.
Danesi, M. (2023b). *Poetic logic and the origins of the mathematical imagination*. Springer.
Descartes, R. (1637). *La géometrie*. Presses Universitaires de France.
Devlin, K. (2011). *The man of numbers: Fibonacci's arithmetic revolution*. Walker and Company.
Drake, M. (2014). Learning to measure length: The problem with the school ruler. *Australian Primary Mathematics Classroom, 19*, 27–32.
Fauconnier, G., & Turner, M. (1998). Conceptual integration networks. *Cognitive Science, 22*, 133–187.
Fauconnier, G., & Turner, M. (2002). *The way we think: Conceptual blending and the mind's hidden complexities*. Basic.
Fauconnier, G., & Turner, M. (2003). Conceptual blending, form and meaning. *Recherches en Communication, 19*, 48413. https://doi.org/10.14428/rec.v19i19.48413
Fibonacci, L. (1202). *Liber Abaci (Abbbaci)*. Internet Archive. https://archive.org/details/LiberAbaci
Fortnow, L. (2013). *The Golden ticket: P, NP, and the search for the impossible*. Princeton University Press.
Frykholm, J. (2010). *Learning to think mathematically with the number line*. Cloudbreak Publishing.
Gödel, K. (1931). Über formal unentscheidbare Sätze der Principia Mathematica und verwandter Systeme, Teil I. *Monatshefte für Mathematik und Physik, 38*, 173–189.
Guhe, M., Pease, A., Smaill, A., Martinez, M., Schmidt, M., Gust, H., Kühnberger, K.-U., & Krumnack, U. (2011). A computational account of conceptual blending in basic mathematics. *Cognitive Systems Research, 12*, 249–265.
Hilbert, D. (1902). Mathematical problems. *Bulletin of the American Mathematical Society, 8*, 437–479.
Hilbert, D. (1926). Über das Unendliche. *Mathematische Annalen, 95*, 161–190.
Hilbert, D. (1928). Die Grundlagen der Mathematik. *Abhandlungen aus dem Seminar der Hamburgischen Universität, 6*, 65–85.
Hofstadter, D. (1979). *Gödel, Escher, Bach: An eternal Golden Braid*. Basic Books.
Hofstadter, D. (2007). *I am a strange loop*. Basic Books.
Johnson, M. (1987). *The body in the mind: The bodily basis of meaning, imagination, and reason*. University of Chicago.
Kaplan, R. (2000). *The nothing that is: A natural history of zero*. Oxford University Press.
Kunen, K. (1980). *Axiomatic set theory*. Elsevier.
Kutter, E. F., Bostroem, J., Elger, C. E., Nieder, A., & Mormann, F. (2022). Neuronal codes for arithmetic rule processing in the human brain. *Current Biology, 32*, 1275–1284.e4. https://doi.org/10.1016/j.cub.2022.01.054
Lakoff, G. (1979). The contemporary theory of metaphor. In A. Ortony (Ed.), *Metaphor and thought* (pp. 202–251). Cambridge University Press.
Lakoff, G. (1987a). *Women, fire and dangerous things: What categories reveal about the mind*. University of Chicago Press.
Lakoff, G. (1987b). Cognitive models and prototype theory. In U. Neisser (Ed.), *Concepts and conceptual development: Ecological and intellectual factors in categorization* (pp. 63–100). Cambridge University Press.
Lakoff, G. (2014). Mapping the brain's metaphor circuitry: Metaphorical thought in everyday reason. *Frontiers in Human Neuroscience, 8*, 958. https://doi.org/10.3389/fnhum.2014.00958
Lakoff, G., & Johnson, M. (1980). *Metaphors we live by*. University of Chicago Press.
Lakoff, G., & Núñez, R. (2000). *Where mathematics comes from: How the embodied mind brings mathematics into being*. Basic Books.

References

Maas, C. U. (2023). By spending some time noticing and talking about the number zero, children can develop their understanding of this key math concept. *Development and Research in Early Mathematics Education.* https://dreme.stanford.edu/news/exploring-the-number-zero-helping-children-understand-the-empty-set/

Peano, G. (1908). *Formulario Mathematico.* Bocca Frères.

Peirce, C. S. (1931–1958). *Collected papers of Charles Sanders Peirce* (Vols. 1–8), C. Hartshorne & P. Weiss (Eds.). Harvard University Press.

Poincaré, J. H. (1902). *Science and hypothesis.* Dover.

Posamentier, A. S., & Lehmann, I. (2007). *The (fabulous) Fibonacci numbers.* Prometheus.

Reddy, M. J. (1979). The conduit metaphor: A case of frame conflict in our language about language. In A. Ortony (Ed.), *Metaphor and thought* (pp. 284–310). Cambridge University Press.

Richards, I. A. (1936). *The philosophy of rhetoric.* Oxford University Press.

Seife, C. (2000). *Zero: The biography of a dangerous idea.* Penguin.

Soskice, J. M. (1985). *Metaphor and religious language.* Clarendon Press.

Susac, A., & Braeutigam, S. (2014). A case for neuroscience in mathematics education. *Frontiers in Human Neuroscience, 8,* 314. https://doi.org/10.3389/fnhum.2014.00314

Tao, T. (2019). *The notorious Collatz conjecture.* terrytao.files.wordpress.com/2020/02/collatz

Terras, R. (1976). A stopping time problem on the positive integers. *Acta Arithmetica, 30,* 241–252.

Thom, R. (1975). *Structural stability and morphogenesis: An outline of a general theory of models.* Benjamin.

Thom, R. (2010). Mathematics. In T. A. Sebeok & M. Danesi (Eds.), *Encyclopedic dictionary of semiotics* (3rd ed.). Mouton de Gruyter.

Tsang, J. M., Blair, K. P., Bofferding, L., & Schwartz, D. L. (2015). Learning to 'see' less than nothing: Putting perceptual skills to work for learning numerical structure. *Cognition and Instruction, 30,* 154–197.

Turing, A. (1936). On computable numbers with an application to the Entscheidungs problem. *Proceedings of the London Mathematical Society, 41,* 230–265.

Turner, M. (1997). *The literary mind.* Oxford University Press.

Turner, M. (2005). *Mathematics and narrative.* thalesandfriends.org/en/papers/pdf/turner.paper.pdf

Turner, M. (2012). Mental packing and unpacking in mathematics. In M. Bockarova, M. Danesi, & R. Núñez (Eds.), *Semiotic and cognitive science essays on the nature of mathematics* (pp. 123–134). Lincom Europa.

Turner, M. (2014). *The origin of ideas: Blending, creativity, and the human spark.* Oxford University Press.

Veale, T., & Keane, M. T. (1992). Conceptual scaffolding: A spatially founded meaning representation for metaphor comprehension. *Computational Intelligence, 8,* 494–519.

Yong, L. L. (1994). Jiu Zhang Suanshu 九章算術 (Nine chapters on the mathematical art): An overview. *Archive for History of Exact Sciences, 47,* 1–51.

Chapter 4
Learning, Diagrams, and AI

4.1 Introduction

Questions about the psychological validity of the notion of image schema have surfaced ever since it crystallized within the cognitive sciences in the 1980s as a model of how meaning unfolds in language and subsequently utilized by Lakoff and Núñez in *Where Mathematics Comes From* (2000) to explicate how it informs math cognition in its essence. Concretely, the critiques have revolved around the following key question: Is there any empirical evidence that image schemas are "real" in any psychological or neurological sense? This chapter addresses this question from three angles:

1. It discusses the relevant psychological and educational research on how mathematics is learned in childhood, which seems to provide substantive support for the notion of image schemas.
2. It then examines the forms and functions of diagrams in mathematics because, as easily examinable "drawings," they can be used inferentially as "external models" of the inner (neural) features of image-schematic thought.
3. Finally, the research being carried out on image schemas within AI is discussed as further corroboration of the soundness of the notion.

This chapter ends with an overall assessment of the psychological validity of image schema theory. Adopting this theory and the processes that it enfolds or subtends (Chap. 3) might, in fact, be nothing more than imaginative speculation about how mathematics unfolds in the brain, and how it becomes embedded as a knowledge system in the world. Given this real possibility, it is little wonder that Lakoff (2009) himself reviewed relevant work in neuroscience a few years after the publication of *Where Mathematics Come From*, aiming to provide psychological substance to both the notion of conceptual metaphor and its basis in image schema theory (Lakoff, 2008, 2009, 2014). In his literature review, Lakoff claimed to have found corroborative neurological evidence that the conceptual system he describes in his own work is acquired "via neural learning mechanisms early in life...just by functioning in the everyday world." This is apparently accomplished, as he goes on to observe, by a process whereby "Each primary metaphor neurally maps one primitive schema onto another, creating an asymmetric circuit linking them."

The research Lakoff describes, albeit highly selective, does indeed seem to suggest that image schemas can be adopted, at the very least, to account for the documented fact that children learn to speak and count at the same time, beyond instinctive counting, and with the same kinds of verbal expressions and gestural-physical behaviors. Clearly, the empirical work conducted on mathematical learning in children (both in the past and in the present) can be employed as a litmus test for assessing the explanatory plausibility of image schema theory in order to gauge if it fits the data, and if not, how divergent is it from the facts. Another key area of analysis, which bears substantial relevance to the topic at hand is, as mentioned above, the examination of the diagrams that have been used, and continue to be used, to model mathematical structures, ideas, and processes. Diagrams can be defined, for the present purpose, as drawings revealing the form of inner image-schematic thinking. The same confirmatory approach can be discerned in the use of AI to model image schemas. The diagrammatic form in this case is to be found in algorithms, whose form can be used to assess the validity of the notion.

Another area for litmus-testing image schema theory, albeit indirectly, is in mathematics education—an area to which the theory has been applied most fruitfully, adopted by many teachers as implying a set of pedagogical ideas (for example, Presmeg, 1997, 2005; Yee, 2017). Children learn best by experiencing the meaning of new concepts through the senses and the body. Manipulatives, for instance, have been used commonly to impart the concepts of quantity and numeration, ever since Maria Montessori introduced them formally into childhood education in the first decades of the twentieth century (Montessori, 1909, 1916). A manipulative is designed to get learners to literally grasp a conceptual distinction, such as *larger* versus *greater* via the manipulation of objects of varying sizes or quantities—reflecting the container schema in its essence.

4.2 Learning

Instinctive counting (number sense), subitizing, and spatiality emerge during the "pre-math" phase of development, as argued throughout this book. These are innate skills that human infants share with other species (Gelman & Gallistel, 1978). Moreover, they surface in tandem, as Kaufman et al. (1949) noted with regard to the subitizing skill, finding that young children can instinctively group the elements consisting of four to five items within a certain visual scene—a skill that implies overlapping number sense with spatiality cognitive processes. There is little doubt that humans are born with an innate number sense, as a plethora of evidence strongly indicates (Beck & Clarke, 2023), even though there is some debate as to what "number" means in this developmental scenario. As Núñez (2017) observes, a dissenting voice in the debate, the term "number sense" cannot possibly refer to numbers, because numbers are precise: 10 is exactly one more than 9 and exactly one less than 11. So, Núñez argues, number sense is nothing more than a synonym for subitizing.

Whatever the truth of the matter, it is during the "pre-math" phase that these innate skills gradually develop into arithmetical and geometrical concepts guided by the same type of image-schematic thought system that undergirds language and other faculties. The fact that language is used to impart elementary arithmetical concepts to children suggests that number and word sense are intertwined. It is during the "pre-math" and "math" phases that the first primitive image schemas manifest themselves simultaneously in the child's speech and counting activities. These can be seen as operative in the content of the inner speech that children display as they "discuss to themselves" the ideational patterns that they perceive in their everyday situations (Vygotsky, 1962). Across languages and cultures, the container and path schemas show up in the arithmetical conceptualizations that have been recorded in the inner speech of children (Alderson-Day & Fernyhough, 2015; Munroe, 2023). As they speak to themselves they put things into piles according to size or some other distinctive feature (such as shape), indicating a sense of containment as a classificatory conceptual system—a behavior documented as well by Piaget (for example, 1941). Children might also use the length of an array of objects laid out before them to conceptualize the number of objects in the array—the longer of two arrays is perceived by the child as having the greater number of objects. Sorting, piling, and lining up objects constitute behaviors that suggest the operation of first-order image schemas in guiding the child's understanding of mathematical pattern.

Pedagogy in elementary school mathematics typically reflects an intuitive sense of utilizing the conceptual metaphors that result from these schemas with instructions or explications such as: "adding two numbers is like putting objects in two hands together," revealing a reference to hands as containers. As Devlin (2011, 2013) has amply documented, without such instruction, learning difficulties might emerge or else may be confused by the inconsistency of the instructional language used to describe mathematical ideas (see also Scheiner et al., 2022). Consider a common technique used to teach fractions to young children in early grades—a method based on a blending of the part-whole and partitioning image schemas translated concretely into pedagogical diagrams such as the following ones (for example, Mighton, 2019) (Fig. 4.1).

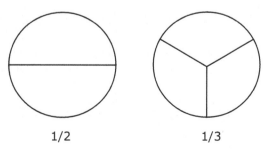

Fig. 4.1 Teaching fractions with image-schematic diagrams

The idea is to portray an object, such as a round pie, as "fractionable." So, the child may be instructed as follows: If you cut the pie in half, you will get two equal pieces. Each of these is "one of two" or in numeral form, 1/2. If you cut the same pie into three equal parts, you will produce three equal pieces. Each of these is "one of three" or 1/3. Now the same diagrammatic pedagogy can be employed to compare the value of the two fractions, since it can be seen concretely via the two diagrams above that 1/2 represents a larger piece of the pie than does 1/3. Thus, the fraction 1/2 is greater in value than 1/3. The diagrams constitute visual mappings of the part-whole and partitioning image schemas. In a comprehensive study, analyzing the use of such pedagogical techniques, Monson et al. (2020) found that the circle-pie type of diagram significantly enhanced the child's learning of the fraction concept, more so than any other pedagogical technique, concluding that: "fraction circles provide students with the opportunity to develop a flexible interpretation of what a unit can be, which is important when students are asked to order 2 fractions" (Monson et al., 2020: 119).

The question becomes: How do these intuitive pedagogical constructs relate to number sense, if at all? As mentioned briefly in Chap. 1, Brian Butterworth (1999) envisions number sense as an evolutionary mechanism specialized for extracting quantitative information from the environment, organized in the brain differentially from the way word sense is. The non-linguistic nature of elemental mathematics, according to Butterworth, would explain why cultures that have no symbols or words for numbers have still managed to develop counting systems for practical purposes. While this is certainly correct, it appears to put aside the difference between numbers and numerals. It is arguably more accurate to say that there are cultures without numerals (as we understand them in standard mathematical tradition), or with only a few signs representing numbers. One of these is the Pirahã culture in Amazonia (Everett, 2005), which does not need precise numeration systems for its everyday activities, and thus making do with words such as "few" or "some." In effect, such cultures have subitizing competence but do not refine it into precise numeration skill with words and numerals in the same way as in mathematical traditions. This does not block their arithmetical competence, however, since it is a sophisticated self-contained one that parallels the traditions.

As also discussed in Chap. 1, Stanislas Dehaene (1997) similarly posits the existence of separate neural regions for language and mathematics. By tracing the history of numbers, from prehistoric times when people, as Dehaene suggests, indicated a number by pointing to a part of their body, to modern numerals and number systems, the evidence suggests, in his view, parallel but autonomous developments of number and word sense. But this too appears to avoid some aspects of evolution that might come into play in the overall neurological scenario. It was the advent of bipedalism on the human evolutionary timetable (during the *homo erectus* era) that liberated the fingers to do several things—to count and gesticulate. Gesture has been shown to constitute a proto-form of language, having left its residues in gesture languages and in the gesticulations that accompany oral speech. The fact that gesture becomes a default form of communication among speakers of different languages is indirect evidence of this evolutionary event, suggesting more broadly that number and word sense (based on gesture) were co-occurrent in the evolutionary scenario.

Research by Gunderson et al. (2015) would seem to contradict my latter assertion, since the researchers found that preschoolers showed number knowledge in gestural forms (such as counting with fingers) that is not matched by words in their speech, thus seemingly supporting the theory of separate brain modules for number and word sense. However, it is not possible to interpret early number gestures as cognitively autonomous signs, since they occur typically as speech-based sign behaviors. Moreover, contrasting evidence was brought forth by Wiese (2007) which showed that the only way number sense "could evolve in humans is via verbal sequences that are employed as numerical tools, that is, sequences of words whose elements are associated with empirical objects in number assignments." Relevant neuroscientific research, reviewed by Tim Rohrer (2005), suggests in fact that number and word sense are not locatable in strictly differentiated neural regions but are spread throughout the brain, indicating that they are likely intertwined conceptually. Without going further into the debate, suffice it to say that the evidence for the evolution of separate number and word senses is not as clear-cut as it is made out to be.

One specific way to investigate the psychological plausibility of image schema theory as a framework to examine difficulties associated with math learning in formal educational contexts, a phase when number sense is developing into "math" sense. In 2003, the present author used this framework to teach algebraic story problems in a course taken by self-defined "math phobics" at the University of Toronto, all of whom were native speakers of English. The study produced interesting and relevant results that became the impetus for a follow-up study conducted in early 2007 (Danesi, 2007) in which the same pedagogical approach to teaching story problems in an elementary school setting was used, thus allowing for a re-assessment of the outcome to the original 2003 case study. To see what was involved pedagogically, consider the following story problem that was used in the experiment:

John is five years older than Mary. Four years from now, he will be twice her age. What is the present age of each?

An image-schematic analysis of this problem shows that it is constructed on the basis of two image schemas of *time*, which are indexed by the differential use of the prepositions *since* and *for* in common sentences such as the following:

- I have been living here *since* 2000.
- I have known Lucy *since* November.
- I have not been able to sleep *since* Monday.
- I have been living here *for* 25 years.
- I have known Lucy *for* 9 months.
- I have not been able to sleep *for* 5 days.

An analysis of the complements that come after *since* reveals that they are metaphorical mappings of a source domain based on the path image schema, whereby "time" is conceptualized as a point on a timeline: *2000, November, Monday*. In contrast, the complements that follow *for* reflect a concept of time as a container with time units as numerical quantities within it: *25 years, 9 months, 5 days*. In other words, the problem is embedded in two conceptual metaphors—*time is a point on a timeline* and *time is a container of numerical quantities*, reflecting two first-order image schemas—the path and container schemas. The "math phobics" course, which I taught myself, was an offering of the School of Continuing Studies at the University of Toronto. At the start of the course, 19 of the 25 adults who enrolled in it had clear difficulties understanding story problems algebraically. As one person put it: "I have no clue how to do this." So, I explicitly taught the class how to decode the conceptual structure of such problems—that is, I showed them how to unpack the conceptual metaphors that constituted them.

The pedagogical method is summarized as follows: The students had taken algebra in high school and were thus familiar with the technique of using letters for unknowns. So, I explained, if we let x represent Mary's present age, then John's present age would be represented by $x + 5$. Why, I was asked by a few of the students? This is so, I clarified, because the ages are points on a timeline, on which John's age-point is "5 points" to the right of Mary's age-point. Now, "4 years from now" involves movement along the same timeline—it means moving John's and Mary's individual age-points "to the right by 4" on it, with the present being the zero starting point ("now"): algebraically, this translates as $x + 4$ for Mary and $x + 5 + 4 = x + 9$ for John. The diagram below was used as a visual representation to represent the conceptual metaphor (Fig. 4.2).

Finally, to set up an algebraic relation, it was explained to the students that it was necessary to "shift conceptual metaphors," that is, to the *time is a container of*

Mary's age four years from now, shown on a year-by-year basis

Timeline	0	1	2	3	4
	↓	↓	↓	↓	↓
Representation	x	x + 1	x + 2	x + 3	x + 4
	↓	↓	↓	↓	↓
Meaning	age now	one year from now	two years from now	three years from now	four years from now

John's age four years from now, also shown on a year-by-year basis

Timeline	0	1	2	3	4
	↓	↓	↓	↓	↓
Representation	x + 5	x + 5 + 1 = x + 6	x + 5 + 2 = x + 7	x + 5 + 3 = x + 8	x + 5 + 4 = x + 9
	↓	↓	↓	↓	↓
Meaning	age now	one year from now	two years from now	three years from now	four years from now

Fig. 4.2 Image-schematic pedagogical diagram

numerical quantities one. John's age is quantifiable as twice that of Mary's age. This is conceptually a difference in the size of the container schemas that hold the two ages (conceptualized as entities). This step entails expressing in algebraic notation the comparison between the two containers. We have established that John's age 4 years from now is $(x + 9)$. We also know that this is twice Mary's age at that time, that is $(x + 4) + (x + 4)$, or $(2x + 8)$. The two are equal in quantity, so: $(x + 9) = (2x + 8)$. Solving the equation, which the students knew how to do, produces the solution of $x = 1$, which means that Mary is 1 year old and John, who is 5 years older, is thus 6 years old. Every story problem used in the course was explained in an analogous way. By the end of the course, 23 of the 25 students claimed to have grasped the principles underlying the solutions to such story problems. This result, albeit anecdotal and based on a small sample of self-proclaimed math phobics, nonetheless was suggestive of two things: (a) image-schematic pedagogy is simple to understand and (b) it seems to allow learners to access the underlying conceptual structure of algebraic story problems.

The follow-up 2007 study involved grade eight students in the city of Toronto. Several research assistants were hired at the University of Toronto, who were trained in advance in image schema pedagogy and told to use whatever insights it afforded them individually to prepare learning materials and techniques (illustrative problems, diagrams, etc.) as required. At the time, grade eight was the point in the Ontario math curriculum where the solution of story problems at different levels of difficulty was considered to be a required skill. A number of grade eight math teachers in local Toronto schools participated in the project by helping the research team identify the students experiencing the most difficulty in solving such problems. Sixty-four such students were identified to the team. They were then given a test consisting of typical grade eight level story problems (with permission of the parents and the principals of the schools). This allowed the team to ascertain that 49 students were candidates for the project.

Each of the students met with a member of the research team for 15 minutes after school hours on a specified day of the week. The same story problems used in their regular classrooms were taught again to each student using image schema pedagogy: that is, each student was trained how to decode the conceptual metaphors that undergirded the problems, in similar ways to the 2003 project (above). The progress of each student was charted on a regular basis, using the marking system in place at the time in Ontario, based on a percentage grade out of 100. At the end of the school year, the scores that each student obtained in story problem-solving in class initially and then after a period of tutoring with the research assistants were compiled and assessed. The cumulative average score for the group of 49 students went up from around 20% at the start to 82% (standard deviation of 2.3%). Although such positive results could be attributed simply to the fact that the students received extra individualized attention, they nevertheless provided supportive evidence to the 2003 study that image schema pedagogy appeared to enhance learning in this problem-solving area. Moreover, the two studies indicated that age was not a factor, given that the 2003 study involved adults while the 2007 study involved elementary school children.

Similar findings have come forth from other studies utilizing image-schematic-based methods of teaching, either explicitly or implicitly (for example, Berggren, 2022; Danesi, 2023; David, 2020; English, 1997; Hughes & Cuevas, 2020; Williams, 2019; Yee, 2017). In a literature review of 70 such studies, Falani et al. (2023) concluded that there was a consistency of positive outcomes in teaching mathematics at various school levels with the use of methods based on conceptual metaphor their, suggesting that image schemas are hardly figments of the theorist's mind, but rather actual learning-enhancing mechanisms that can be activated profitably by educators to assist students in learning a vast array of concepts.

4.3 Diagrams

Image schema theory is ultimately based on the premise that the brain produces imagistic mental structures on the basis of experiential information inputs, which mirror physical experiences in a cognitive outline form—paths, journeys, containers, partitioning, connecting, collating, etc. These outline mind forms come out, often, in the actual physical form of diagrams, such as the timeline one above, the classic number line diagram, the Eulerian graph network diagram, and so on, which can be assumed to resemble the form that image schemas might take in the mind, with variation in detail, of course. To put it another way, diagrams are visual pencil-and-paper models of image-schematic structures in the mind. The term "model" is used somewhat liberally here with respect to how it used more generally in psychology (Craik, 1943; Johnson-Laird, 1983). A diagram is a visual model drawn by either a person or a programming language which can be used as an analogue model of the path image schema.

Research on metaphor processing has actually shown a mirror relation between mental imagery and drawings. In 1975, for instance, Billow found that child subjects, from 5 to 13 years of age, could easily give a pictorial form to metaphorical utterances, implying that they were able to flesh out the formative image schemas embedded in them. Since the use of prompts did not significantly improve the sketching process or the time required to draw the metaphors, Billow concluded that children can easily and uniformly imagine the mental imagery associated with metaphorical thought and draw it on paper. Research by Kennedy (1984, 1993, 2020) on congenitally blind people has shown that they possess the same kind of pictorial modeling ability that visually normal people have; that is, they are capable of making appropriate line drawings of metaphorical concepts if given suitable contextual information and prompts. This strongly suggests that the image-schematic structure of metaphors is the result of an innate imagistic mechanism.

On the basis of such research, it can be argued that diagrams such as the timeline one above are pedagogically effective because they tap into the students' inner image-schematic thought processes, and this is why they are not only highly understandable, no matter the student's age, but also inducive of relevant mathematical ideation in a specific area of learning (such as story problems). This might also

4.3 Diagrams

explain, more broadly, why diagrams permeate mathematics, for illustrating theorems, conducting proofs, and even founding a new branch. Diagrammatic creations such as those devised by Euler or Cantor are concrete examples of how mathematicians represent image-schematic concepts in pictorial form. They are "drawings" of the image schemas in their brains, and since we understand them as well, the implication is that they are real in a psychological sense.

Consider graph theory once again. As discussed, it originated with the Königsberg Bridges Problem, which inspired Euler to devise a diagram that mirrored in visual outline the physical path network of the bridges, with vertices and edges. After solving the problem with the aid of his diagram, Euler foresaw its implications more generally, leading to his establishment and further development of graph theory. The mappings of this first-order diagrammatic model (that is, a diagram representing a first-order image schema in outline pictorial form) onto other target domains have made possible many other conceptual forms. For example, one might ask: What is the shortest path from "a" to "h" in the network diagram below, if it is necessary to visit each node.

The minimal path is (a–b–c–d–e–f–g–h). Now, if we were asked to find a path in which it is necessary to visit each node in a network once and only once, without doubling back on any route, then the problem is turned into a famous one in graph theory, known as the *traveling salesman problem*, paraphrased below in its simplest terms (Benjamin et al., 2015: 122):

> A salesperson wishes to make a round-trip that visits a certain number of cities. The person knows the distance between all the cities. If each city must be visited exactly once, then what is the minimum total distance of such a round trip?

In the network represented by (Fig. 4.3), there is no solution to the problem, since, as in Euler's Königsberg Bridges Problem, each node cannot be traversed once and only once without doubling back at a node. Modifying it, however, produces a network in which it is now possible to visit every node exactly once to make a complete round trip (Fig. 4.4)—the distances between the nodes (cities) are called weights.

A complete round trip through the planar network in which each node is visited once and only once is the following one: a–b–c–d–e–f–g–h–a. The total length is:

Fig. 4.3 A network diagram

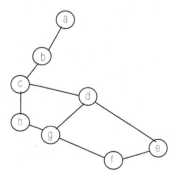

Fig. 4.4 A network for the traveling salesman problem

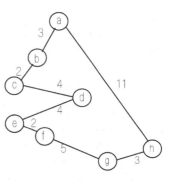

3 + 2 + 4 + 4 + 2 + 5 + 3 + 11 = 34 units. This type of network is called, more exactly, a Hamiltonian cycle. More technically, the problem is to find a minimum weight summation in the Hamiltonian cycle. Now, solving the problem may appear to be simple task, consisting of the following solution strategy: calculate the weights to transverse all possible routes and select the shortest. However, as it turns out, this is extremely time consuming and as the number of cities grows, whereby the solution becomes increasingly intractable. A network with just 10 cities has 9! or 362,880 possible routes, which is too many even for a computer to handle in a reasonable time. In effect, the traveling salesman problem is a so-called NP-hard problem for which there is no polynomial time algorithm that is known to efficiently solve it.

The point here is that the first-order Eulerian diagram, based on a connected path image schema, has led, through elaborations and extrapolations, not only to a branch of mathematics but also to a central problem in computer science (Flovik, 2022; Newman, 2010). No algorithm exists that would guarantee the shortest path through any Hamiltonian network. The time a computer takes to solve a problem is represented as a polynomial function of the size of the input. The expression "quickly solvable" means that an algorithm can solve a task in polynomial time. The general class of problems for which an algorithm can provide an answer in polynomial time is represented with "P," for "polynomial time." The class of problems to which there is no known way to find an answer quickly, although still verifiable in polynomial time, is represented with "NP," which stands for "nondeterministic polynomial time." The "P versus NP" problem is one of the most important open problems in computer science.

The formal articulation of the problem is traced to a 1971 paper by computer scientist Stephen Cook and, independently, to another paper by Russian computer scientist Leonid Levin, in 1973. An early mention of the problem is found in a 1956 letter written by Kurt Gödel to John von Neumann. Gödel asked Neumann whether theorem-proving could be solved in quadratic or linear time (Hartmanis, 1989). From the letter, the following system of problem-solving emerged: "P" would consist of all those problems that can be solved on a deterministic sequential machine in an amount of time that is polynomial; the class "NP" would consist of all those problems whose solutions can be verified in polynomial time given the relevant information, or equivalently, whose solution can be found in polynomial time on a nondeterministic machine. If "P = NP," every "NP" problem would contain a shortcut, allowing computers to quickly find solutions to them. But if "P ≠ NP," then no such shortcuts exist, and the problem-solving powers of computers will remain permanently limited. Some "P" problems are also in "NP." But the really hard ones are only in "NP," called "NP-complete." Without going further into details here, suffice it to say that such problems would not have been thinkable without the development of the notions of graphs and networks, as models of computation, which have their ideational source in first-order image-schematic cognition.

Truly understanding mathematics implies grasping how mathematicians use modeling strategies such as diagrams to carry out their activities. Path diagrams are cases-in-point. The origination process might go somewhat as follows. First, a "metaphorical hunch" is guided by a first-order image schema, which often inspires the ideation of a diagram, such as the graph created by Euler for the Königsberg Bridges network. Once devised, the diagram leads, second, to the revelation of properties of a network or other system that would not have been obvious otherwise. Third, it then becomes a model of mathematical structure in a certain domain, becoming the basis of a new branch—graph theory (Fig. 4.5).

Expressed in language, as was the case with the Königsberg Bridges Problem initially, we would literally not be able to *see* the possibilities that a diagram presents to us through its image-schematic structure. To use Susan Langer's (1948) theory of representation, it can be said that a diagram is a presentational form—that is, it tells us much more than a statement does because it literally "presents" the image-schematic structure inherent in something. We do not read a diagram or a melody, she emphasized, as made up of individual bits and pieces, but as a totality whose meaning is greater than the sum of the parts. In other words, diagrams "show" relations in a holistic way that are not apparent in language descriptions.

Path problems are, in sum, those that require us to identify the optimal sequence of stages or paths needed for resolving a given situation. They stem ultimately from practical experiences: What is the best way to cross over seven connected bridges without doubling back on any of them? What is the shortest and best route connecting various points? How can I cross a river with a wolf, a goat, and a head of cabbage with a boat that holds only two (the famous river crossing problem of Alcuin)?

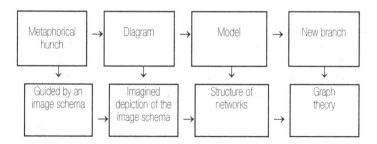

Fig. 4.5 Theory-making process

Fig. 4.6 A simple diagram illustrating the four-color theorem

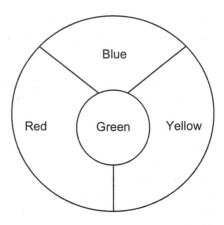

And so on. These experiences tap into first-order image-schematic thinking in the mind, which then guides the mathematician to devise a way to make the structure visible, after which it becomes itself a source for further mathematics, even opening up new vistas (metaphor intended) within mathematics and related fields.

One last example of how the path schema constitutes the source of a particular class of diagrams is the four-color theorem, which is worth discussing in summary form here, since it exemplifies how the same image-schematic structure undergirds many seemingly unrelated areas on the surface. Mapmakers had known since antiquity that four colors were sufficient to color any map, so that no two contiguous regions would share a color. This caught the attention of mathematicians in the nineteenth century, after a young mathematician at University College, London, named Francis Guthrie, formally iterated the mapmaker's knowledge as a problem in 1852. Guthrie apparently wrote about it, first, to his younger brother, Frederick, who, the story then goes, presented it to his own professor, Augustus De Morgan, who quickly realized that the problem had many important ramifications for mathematics. Word of the problem spread quickly. The map below is a simple diagrammatic illustration that four colors will suffice, using blue, red, green and yellow color categories for the sake of concreteness—but a combination of any four colors would do just as well (Wilson, 2002) (Fig. 4.6).

In 1879, Arthur Cayley was the first to envision the problem as a graph-theoretic one by observing that it was possible to solve a version of the problem by limiting the way the boundaries met. So, for example, maps with just three countries have three edges. In the same year, Alfred Kempe employed a similar technique. Without going into details here, Kempe showed that the problem has a connected path structure, subsequently known as a Kempe chain. However, a few years later, Percy Heawood (1890) showed that the proof contained an important error. Nonetheless, the fact that the problem could be conceptualized in terms of graph theory was a significant episode in the history of the problem. In 1880, P. G. Tait established that maps where an even number of boundary lines meet could be colored with no more than four colors, although this result had appeared earlier in Kempe's papers. In 1946, however, Tutte found the first counterexample to Tait's conjecture, leading to the rejection of the proof used. In the twentieth century, mathematicians focused on modifying these kinds of techniques to reduce complicated maps to special cases which could be identified and investigated as to their particular properties. Starting in 1976 at the University of Illinois, Kenneth Appel and Wolfgang Haken eventually reduced the problem to a set of maps with 1936 configurations, checking the maps one by one with different computer programs (Haken & Appel, 1976, 2002). Their method showed that no map with the smallest possible number of regions requiring five colors existed in the set.

Starting with Cayley, as mentioned, a simpler statement of the problem can be formulated in terms of graph theory, whereby the regions of a map can be represented in terms of a planar graph that has a vertex for each region and an edge for every pair of regions that share a boundary segment. In graph-theoretic terminology, the four-color problem can then be formulated in path schematic terms: Can the vertices of every planar graph—a graph in which no edges cross—be colored with at most four colors so that no two adjacent vertices receive the same color, or is "every planar graph four-colorable?" (Wilson, 2002). In this way, as Richeson (2023) aptly puts it: "the map coloring problem becomes a graph coloring problem," and thus the problem can be reformulated as follows: "Prove that the chromatic number of every simple planar graph is at most four." However, no proofs using graph theory to solve the four-color problem have yet to emerge. The point is that this problem is not an isolated one but belongs to the class of problems that fall under the rubric of graph theory.

4.4 AI Modeling

Diagrams using graph theory constitute concrete visual models of image-schematic structure and thus can be given an algorithmic form as a basis for testing the validity of that very structure (Bronstein et al., 2017; Gómez-Ramírez, 2019). Neural network architecture itself is constructed with a graph-theoretic network structure that is based on its design, ultimately, on features associated with the path schema. As such, it provides computational affirmation of the validity of image schema theory.

A neural network possesses the ability to make decisions and acquire knowledge by mimicking the way biological neurons work together to identify concepts in novel information. A neural network consists of the same layered abstract pattern of the connected edges (paths) and nodes (vertices) of mathematical graphs: (1) an input layer, (2) a hidden (in between) layer, and (3) an output layer. Each node is connected to other nodes and the edges can crisscross. The whole network has an associated weight and threshold—a threshold is the value that determines whether an artificial neuron is activated or not. If the output of any individual node is above a specified threshold value, that node is activated, sending data to the next layer of the network. Otherwise, no information is passed along. Weights control the strength of the connection between two nodes, determining how much influence the input will have on the output. Overall, the process of passing data from one layer to the next layer defines a neural network as a connected graph structure (Fig. 4.7).

Fig. 4.7 A simple neural network

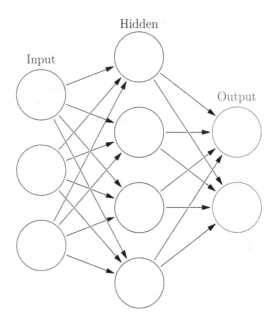

A key computational research question related to a major theme of this book is the following one: Can a neural network be trained to model non-linguistic concepts, such as arithmetical ones, independently of linguistic ones? This question was addressed by Friedemann Pulvermüller (2023) at the Brain Language Laboratory of the Freie Universität of Berlin who examined the role of linguistic conceptual input in the AI modeling of non-linguistic concepts. Pulvermüller and a team of researchers developed a neural network system with artificial neurons that mimicked the activities of the cerebral cortex. The underlying learning principle adopted for the project was Hebbian learning, introduced by Donald Hebb in his 1949 book, *The Organization of Behavior*, which claims that neurons adapt and form stronger connections through repeated use. So, every time a memory is recalled or an action is repeated, the neural pathways involved become more robust as they fire together, making memory easier to reproduce. Applying this principle to the design of their neural network, the team found that interconnected clusters emerged during input training, which activated the relevant concept-forming Hebbian circuitry in the network. They found in particular that non-linguistic concepts were acquired by the network more efficiently and quickly when linguistic inputs were used alongside the non-linguistic inputs. The researchers concluded that the network could better build a representation of an abstract non-linguistic concept if it possessed a linguistic label mapped onto the abstract concept—a finding that has clear implications for the separate-versus-integrated faculty debate, discussed throughout this book.

As Maria Hedblom (2021) has argued and illustrated in several key studies, along with a team of researchers (Hedblom et al., 2017), the mechanisms that guide concept formation in neural networks are describable even at the computational level as first-order image schemas, such as the container and path ones. In one

4.4 AI Modeling

particular study, image-schematic visual representations were used to develop a so-called Diagrammatic Image Schema Language, defined as "a formal representation language that systematizes a set of visual combination rules for different conceptual primitives from the cognitive science literature. These primitives are distinguished from a formal point of view to allow for more general application" (Hedblom et al., 2024).

In another relevant study, Yunus et al. (2022) constructed a procedure that allowed an AI system to learn the vector forms of image schemas from selected input using a base word embedding technique to calculate the final vector representation—in natural language processing, word embedding is the representation of a word as a real-valued vector that positions the word in a vector space consisting of other words that are expected to be similar in meaning. The results of the project give some substance to the clustering structure of image schemas. The researchers describe their project as follows:

> With the image schemas representable as vectors, it also becomes possible to have a notion that some image schemas are closer or more similar to each other than to the others because the distance between the vectors is a proxy of the dissimilarity between the corresponding image schemas. Therefore, after obtaining the vector representation of the image schemas, we calculate the distances between those vectors. Based on these, we create visualizations to illustrate the relative distances between the different image schemas.

In yet another relevant study, Falomir and Plaza (2019) examined computational models of novel concept understanding and creativity from the perspective of image schema theory and conceptual blending theory. In their approach, a novel concept was created by an emitter agent and sent to another agent in the system, which could only process it by blending concepts already known by that agent, implying computationally that the receiving agent of that concept was able to recreate the conceptual network underlying the novel concept. This whole process was guided by the use of image schemas as tools that the agents used to interpret the spatial information obtained when recreating the blend.

As such research indicates, overall, image schema theory and the underlying ideas, processes, and properties behind it have attracted the interest of researchers in AI (Chang et al., 2006; Hedblom et al., 2015, 2016, 2017; Kuhn, 2007). Along with the research on learning mathematics in childhood and at school, as well as the work on diagrams in mathematics, the use of image schema theory in AI constitutes another source of supportive evidence. The overall value of image schemas in AI modeling systems is formulated by Besold et al. (2017: 10), on the basis of the goals of their own research, as follows:

> Image schemas have been proposed as conceptual building blocks corresponding to the hypothesised most fundamental embodied experiences. We formally investigate how combinations of image schemas (or 'image schematic profiles') can model essential aspects of events, and discuss benefits for artificial intelligence and cognitive systems research, in particular concerning the role of such basic events in concept formation. More specifically, as exemplary illustrations and proof of concept the image schemas Object, Contact, and Path are combined to form the events Blockage, Bouncing, and Caused Movement. Additionally, an outline of a proposed conceptual hierarchy of levels of modelling for image schemas and similar cognitive theories is given.

As I read it, this whole line of research is providing two types of findings that support the viability of image schema theory: (1) it gives substance to the idea that diagrammatic representations, such as those discussed in this book, are "resemblance" forms that provide snapshots of the actual mental forms of image schemas and (2) it suggests that some machine languages are based on image-schematic structure, allowing for an algorithmic modeling of this very structure.

4.5 Validation

The studies discussed in this chapter can be seen to provide corroborative support for the initial hypothesis of *Where Mathematics Comes From*—namely, that math cognition is implanted on conceptual metaphorical thinking, which is, in turn, guided by image-schematic processes, along with related processes such as mapping, blending, layering, and clustering. Despite various infelicities that the original theoretical framework might have contained and that the present one is bound to contain as well, at the very least, the notion of image schema cannot be ignored when discussing or researching math cognition. Emerging in early infancy, image schemas enable the formation of abstractions in human thought systems, from language to art and mathematics, showing how they are interrelated cognitively.

An indirect argument in favor of image-schematic cognition involves the reasons for the use of analogy in mathematics and science (Hofstadter & Sander, 2013), which entails the mapping a source domain (the analogue) onto a target domain (a theoretical abstraction), guided by an inherent image schema. Consider a common layperson explication of the theory of relativity, which can be paraphrased as follows. Suppose someone is on a smoothly running train moving at a constant velocity. Even though the train is moving, if a book is dropped or a ball is thrown back and forth, the book will appear to fall straight down and the ball will seem to travel from thrower to catcher as it would outside the moving train. As long as the train runs smoothly, with constant velocity, none of these physical actions will be affected by its motion. Conversely, if the train stops or speeds up abruptly, the ways in which the activities unfold will change. A book may be jarred from a seat and fall without being dropped. The activities involving a thrown ball might be different. Now, the same actions pertain outside the train on *terra firma*—a book will drop straight down, a ball thrown back and forth will move in the same way as it does on a moving train, and so on. From this analogy, it can be seen that the laws of motion are the same in the train as they are on the ground. The principle is as follows: If two systems move uniformly *relative* to each other, then the laws of physics are the same in both systems. Now, the target domain in this case is the notion of relativity and the source domain, used as an analogue, is the movement of a train and the objects in it. The connecting image-schematic link is the operation of the stasis–motion–force properties in the mapping process.

A key critique directed at this overall approach, as encapsulated by Winter and Yoshimi (2020), discussed briefly in Chap. 1, is that the conceptual metaphor

4.5 Validation

approach to math cognition is itself and analogy or a metaphor. As they remark, what the book *Where Mathematics Comes Form* has shown, in the end, is that "abstract thinking is facilitated by metaphor." But then, all theories of mind are analogies and certainly metaphors in some way—there is no theory without a recourse to metaphor, as Max Black convincingly argued in his classic 1962 book, *Models and Metaphors*. Moreover, as discussed in this chapter, evidence of the "reality" of metaphor as an "inner force" for the inducement of mathematical abstractions comes from various areas (learning, diagrams, AI) where image-schematic thought can be seen to be operative in specific ways. As far as can be told, there really are no better ways to explain research in these domains, at least in my view.

Image schema theory is not an all-encompassing theory of math cognition. It provides a snapshot of how some, perhaps many, ideas in mathematics develop initially and evolve sequentially. In their classic work, *The Embodied Mind: Cognitive Science and Human Experience* (Varela et al., 1991), Varela, Rosch, and Thompson argued that "the new sciences of the mind need to enlarge their horizon to encompass both lived human experience and the possibilities for transformation inherent in human experience." Image schema theory falls into an enlarged horizon. The more we probe similarities (or differences) in language and mathematics, with all kinds of research tools, the more we will know about the mind that creates both. The contrasting view that mathematics and language constitute separate faculties, however, cannot be discarded completely. In his relevant book, *Seeing the Mind* (Dehaene, 2023), Stanislas Dehaene describes research he conducted using an MRI scanner to record the brain waves of professional mathematicians as they were presented statements such as "The Euclidean orthogonal group has exactly two connected components" and asked decide whether they were true or false (the statement is true). Other statements were used to examine the brain waves associated with non-mathematical knowledge, as for example, "The end of the Council of Trent coincides with the fall of the Western Roman Empire" (the statement is false).

The results of the study indicated that the brain seems to use different regions to store mathematical and non-mathematical knowledge. When a mathematician thinks about mathematics, the parietal and inferior temporal regions of both hemispheres light up, whereas a completely different set of regions, notably the temporal pole and the angular gyrus, light up when other types of knowledge are being processed. The mathematician's brain therefore seems to house a specialized neural circuitry for mathematical knowledge. Does this mean that mathematicians possess a radically different cortical organization than the average person? The regions for mathematics are present in each of us, Dehaene answers. All it takes to activate them are much simpler statements such as "How much is two times three?"

But there are several things that this overall conclusion seems to skip over. First, it took linguistic stimuli to activate the regions. Would it not make sense to say that verbal circuits synergistically activate math circuits? Is this not the same as saying that they are blended? Second, the AI neural network models discussed in this chapter are strongly suggestive that complex thought involves patterns of connectivity among the nodes in the network, not separate circuitry. Third, it would seem to be more consistent with the observed behaviors of children learning mathematics, even in the "pre-math" stage, to claim that different regions in the brain form interlinked networks in the generation of abstract cognition, even though some may be more active during the processing of specific stimuli, such as those described by Dehaene (Lakoff, 2008; Rohrer, 2005).

It is interesting to note, by way of conclusion, that, even before the advent of conceptual metaphor and image schema theories, the philosopher Ludwig Wittgenstein (1921) had characterized propositions, linguistic and mathematical, as structures designed to represent features of the world in the same way that "pictures" do, echoing a major claim of this book. The lines and shapes of drawings show how things are related to each other; so too, he claimed, do the ways in which words are put together in sentences or symbols in mathematics. Although he did not name his "pictures" as "image schemas," the two seem to be synonymous. Image schema theory formalizes an intuition that we all bear, as the math pedagogy studies discussed in this chapter show—the pictures in our mind reflect pictures in the world, and these lead us to think about the world in highly sophisticated ways.

References

Alderson-Day, B., & Fernyhough, C. (2015). Inner speech: Development, cognitive functions, phenomenology, and neurobiology. *Psychological Bulletin, 141*, 931–965.

Beck, J., & Clarke, S. (2023). Babies are born with an innate number sense. *Scientific American.* https://www.scientificamerican.com/article/babies-are-born-with-an-innate-number-sense/

Benjamin, A., Chartrand, G., & Zhang, P. (2015). *The fascinating world of graph theory*. Princeton University Press.

Berggren, J. (2022). Some conceptual metaphors for rational numbers as fractions in Swedish mathematics textbooks for elementary education. *Scandinavian Journal of Educational Research, 67*, 914–927.

Besold, T. R., Hedblom, M. M., & Kutz, O. (2017). A narrative in three acts: Using combinations of image schemas to model events. *Biologically Inspired Cognitive Architectures, 19*, 10–20.

Billow, R. M. (1975). A cognitive developmental study of metaphor comprehension. *Developmental Psychology, 11*, 415–423.

Black, M. (1962). *Models and metaphors*. Cornell University Press.

Bronstein, M. M., Bruna, J., LeCun, Y., Szlam, A., & Vandergheynst, P. (2017). Geometric deep learning: Going beyond Euclidean data. *IEEE Signal Processing Magazine, 34*, 18–42.

Butterworth, B. (1999). *What counts: How every brain is hardwired for math*. Free Press.

Cayley, A. (1879). On the colourings of maps. *Proceedings of the Royal Geographical Society, 1*, 259–261.

Chang, Y.-H., Cohen, P. R., Morrison, C. T., Amant, R. S., & Beal, C. (2006). Piagetian adaptation meets image schemas: The Jean system. *From Animals to Animats, 9*, 369–380.

Cook, S. (1971). The complexity of theorem proving procedures. *Proceedings of the Third Annual ACM Symposium on Theory of Computing*, 151–158. https://doi.org/10.1145/800157.805047

Craik, K. J. W. (1943). *The nature of explanation*. Cambridge University Press.

Danesi, M. (1987). Formal mother-tongue training and the learning of mathematics in elementary school: An observational note on the Brussels Foyer Project. *Scientia Paedogogica Experimentalis, 24*, 313–320.

References

Danesi, M. (2003). Conceptual metaphor theory and the teaching of mathematics: Findings of a pilot project. *Semiotica, 145*, 71–83.

Danesi, M. (2007). A conceptual metaphor framework for the teaching mathematics. *Studies in Philosophy and Education, 26*, 225–236.

Danesi, M. (2020). *Pi (π) in nature, art, and culture: Geometry as a hermeneutic science*. Brill.

Danesi, M. (Ed.). (2023). *Handbook of cognitive mathematics*. Springer.

David, E. K. (2020). *A comparison of frameworks for conceptualizing graphs in the Cartesian coordinate system*. Mathematica Association of America. http://sigmaa.maa.org/rume/crume2019/Papers/154.pdf

Dehaene, S. (1997). *The Number Sense: How the Mind Creates Mathematics*. Oxford: Oxford University Press.

Dehaene, S. (2023). *Seeing the mind*. MIT Press.

Devlin, K. (2000). *The math gene: How mathematical thinking evolved and why numbers are like gossip*. Basic.

Devlin, K. (2005). *The math instinct*. Thunder's Mouth Press.

Devlin, K. (2011). *Mathematics education for a new era: Video games as a medium for learning*. CRC Press.

Devlin, K. (2013). The symbol barrier to mathematics learning. In M. Bockarova, M. Danesi, & R. Núñez (Eds.), *Semiotic and cognitive science essays on the nature of mathematics* (pp. 54–60). Lincom Europa.

English, L. D. (Ed.). (1997). *Mathematical reasoning: Analogies, metaphors, and images*. Lawrence Erlbaum Associates.

Everett, D. (2005). Cultural constraints on grammar and cognition in Pirahã. *Current Anthropology, 46*, 621–624.

Falani, I., Supriyati, Y., Marzal, J., Iriani, D., & Simatupang, G. M. (2023). Metaphor studies investigation in mathematics education: A systematic review. *Indonesian Journal of Mathematics Education, 6*, 40–62.

Falomir, Z., & Plaza, E. (2019). Towards a model of creative understanding: Deconstructing and recreating conceptual blends using image schemas and qualitative spatial descriptors. *Annals of Mathematics and Artificial Intelligence, 88*, 457–477. https://doi.org/10.1007/s10472-019-09619-9

Flovik, V. (2022). An introduction to graph theory. *Built-in*. https://builtin.com/machine-learning/graph-theory

Gelman, R., & Gallistel, C. R. (1978). *The child's understanding of number*. Harvard University Press.

Gómez-Ramírez, D. A. J. (2019). *Artificial mathematical intelligence*. Springer.

Gunderson, E. A., Spaepen, E., Gibson, D., Goldin-Meadow, S., & Levine, S. C. (2015). Gesture as a window onto children's number knowledge. *Cognition, 144*, 14–28.

Haken, W., & Appel, K. (1976). The solution of the four-color-map problem. *Scientific American, 237*, 108–121.

Haken, W., & Appel, K. (2002). The four-color problem. In D. Jacquette (Ed.), *Philosophy of mathematics* (pp. 193–208). Blackwell.

Hartmanis, J. (1989). Gödel, von Neumann, and the P = NP problem. *Bulletin of the European Association for Theoretical Computer Science, 38*, 101–107.

Heawood, P. J. (1890). Map-colour theorem. *Quarterly Journal of Mathematics, 24*, 332–338.

Hebb, D. O. (1949). *The organization of behavior*. Wiley.

Hedblom, M. M. (2021). *Image schemas and concept invention: Cognitive, logical, and linguistic investigations*. Springer.

Hedblom, M. M., Kutz, O., & Neuhaus, F. (2015). Choosing the right path: Image schema theory as a foundation for concept invention. *Journal of Artificial General Intelligence, 6*, 21–54.

Hedblom, M. M., Kutz, O., & Neuhaus, F. (2016). Image schemas in computational conceptual blending. *Cognitive Systems Research, 39*, 42–57.

Hedblom, M. M., Kutz, O., Mossakowski, T., & Neuhaus, F. (2017). Between contact and support: Introducing a logic for image schemas and directed movement. In F. Esposito, R. Basili, S. Ferilli, & F. Lisi (Eds.), *Advances in artificial intelligence* (pp. 256–268). Springer.

Hedblom, M. M., Neuhaus, F., & Mossakowski, T. (2024). The diagrammatic image schema language (DISL). *Spatial Cognition & Computation*, 1–38. https://doi.org/10.1080/13875868.2024.2377284

Hofstadter, D., & Sander, E. (2013). *Surfaces and essences: Analogy as the fuel and fire of thinking*. Basic.

Hughes, S., & Cuevas, J. (2020). The effects of schema-based instruction on solving mathematics word problems. *Georgia Educational Researcher*, *17*, 1–50.

Johnson-Laird, P. N. (1983). *Mental models: Towards a cognitive science of language, inference, and consciousness*. Cambridge University Press.

Kaufman, E. L., Lord, M. W., Reese, T. W., & Volkmann, J. (1949). The discrimination of visual number. *American Journal of Psychology*, *62*, 498–525.

Kempe, A. B. (1879). On the geographical problem of the four colours. *American Journal of Mathematics*, *2*, 193–220.

Kennedy, J. M. (1984). *Vision and metaphors*. Toronto Semiotic Circle.

Kennedy, J. M. (1993). *Drawing and the blind: Pictures to touch*. Yale University Press.

Kennedy, J. M. (2020). Metaphor and one-off pictures: Touch and vision. In J. Barnden & A. Gargett (Eds.), *Producing figurative expression: Theoretical, experimental and practical perspectives* (pp. 55–83). John Benjamins Publishing Company.

Kuhn, W. (2007). An image-schematic account of spatial categories spatial information theory. *Lecture Notes in Computer Science*, *4736*, 152–168.

Lakoff, G. (2008). The role of the brain in the metaphorical mathematical cognition. *Behavioral and Brain Sciences*, *31*, 658–659.

Lakoff, G. (2009). The neural theory of metaphor. In R. Gibbs (Ed.), *The metaphor handbook*. Cambridge University Press.

Lakoff, G. (2014). Mapping the brain's metaphor circuitry: Metaphorical thought in everyday reason. *Frontiers in Human Neuroscience*, *8*. https://doi.org/10.3389/fnhum.2014.00958

Lakoff, G., & Núñez, R. (2000). *Where mathematics comes from: How the embodied mind brings mathematics into being*. Basic Books.

Langer, S. (1948). *Philosophy in a new key*. Harvard University Press.

Levin, L. A. (1973). *Problems of Information Transmission*, *9*, 115–116.

Marcus, S. (2012). Mathematics between semiosis and cognition. In M. Bockarova, M. Danesi, & R. Núñez (Eds.), *Semiotic and cognitive science essays on the nature of mathematics* (pp. 98–182). Lincom Europa.

Mighton, J. (2019). Using evidence to close the achievement gap in math. In M. Danesi (Ed.), *Interdisciplinary perspectives on math cognition*. Springer.

Monson, D., Cramer, K., & Ahrendt, S. (2020). Using models to build fractions understanding. *Mathematics Teacher Learning and Teaching PK-12*, *113*, 117–123.

Montessori, M. (1909). *Il Metodo della pedagogia scientifica applicato al'educazione infantile nelle Case dei Bambini*. Casa Editrice S. Lapi.

Montessori, M. (1916). *L'autoeducazione nelle scuole elementari*. Loescher.

Munroe, W. (2023). Thinking through talking to yourself: Inner speech as a vehicle of conscious reasoning. *Philosophical Psychology*, *36*, 292–318.

Newman, M. (2010). *Networks: An introduction*. Oxford University Press.

Núñez, R. (2017). Is there really an evolved capacity for number? *Trends in Cognitive Science*, *21*, 409–424.

Piaget, J. (1952). *The child's conception of number*. Kegan Paul.

Presmeg, N. C. (1997). Reasoning with metaphors and metonymies in mathematics learning. In L. D. English (Ed.), *Mathematical reasoning: Analogies, metaphors, and images* (pp. 267–280). Lawrence Erlbaum.

Presmeg, N. C. (2005). Metaphor and metonymy in processes of semiosis in mathematics education. In J. Lenhard & F. Seeger (Eds.), *Activity and sign* (pp. 105–116). Springer.

Pulvermüller, F. (2023). Neurobiological mechanisms for language, symbols and concepts: Clues from brain-constrained deep neural networks. *Progress in Neurobiology, 230*, 102511.

Richeson, D. S. (2023). The colorful problem that has long frustrated mathematicians. *Quanta Magazine.* https://www.quantamagazine.org/only-computers-can-solve-this-map-coloring-problem-from-the-1800s-20230329/

Rohrer, T. (2005). *From perception to meaning.* Walter de Gruyter.

Scheiner, T., Godino, J. D., Montes, M. A., Pino-Fan, L. R., & Climent, N. (2022). On metaphors in thinking about preparing mathematics for teaching: In memory of José ("Pepé") Carrillo-Yáñez. *Educational Studies in Mathematics, 111*, 253–270.

Tait, P. G. (1880). Remarks on the colourings of maps. *Proceedings of the Royal Society, 10*, 729.

Tutte, W. T. (1946). On Hamilton cycles. *Journal of the London Mathematical Society, 21*, 98–101.

Varela, F., Rosch, E., & Thomson, E. (1991). *The embodied mind: Cognitive science and human experience.* MIT Press.

Vygotsky, L. S. (1962). *Thought and language.* MIT Press.

Wiese, H. (2007). The co-evolution of number concepts and counting words. *Lingua, 117*, 758–772.

Williams, R. (2019). The source-path-goal image schema in gestures for thinking and teaching. *Review of Cognitive Linguistics, 17*, 411–437.

Wilson, R. (2002). *Four colors suffice: How the map problem was solved.* Princeton University Press.

Winter, B., & Yoshimi, J. (2020). Metaphor and the philosophical implications of embodied mathematics. *Frontiers in Psychology, 11*, 569487.

Wittgenstein, L. (1921). *Tractatus logico-philosophicus.* Routledge and Kegan Paul.

Yee, S. P. (2017). Students' and teachers' conceptual metaphors for mathematical problem solving. *School Science and Mathematics, 117*, 146–157.

Yunus, F., Clavel, C., & Pelachaud, C. (2022). Representation learning of image schema. *arXiv:2207.08256.* https://doi.org/10.48550/arXiv.2207.08256

Index

A
Accommodation, 45
Achilles and the Tortoise paradox, 38
AI modeling, 25, 85–88
Analogy, 88, 89
Anthropology, 1, 45
Aristotle, 38
Arithmetic is object collection, 11, 27, 28
Artificial intelligence (AI), 1, 41, 49, 73–90
Assimilation, 45

B
Basic Metaphor of Infinity (BMI), 8–11, 32, 33
Binary numbers, 13, 44
Blending, 12, 33, 49, 53, 54, 60–65, 67, 75, 87, 88
Blockage schema, 33
Butterworth, B., 1, 14, 76

C
Calculus, 26, 35, 39, 52, 57, 60
Cantor, G., 9–11, 23, 29, 42, 60, 61, 81
Cardinality, 9, 23, 42, 64
Closure, 35, 36
Clustering, 49, 50, 65–67, 87, 88
Collatz's conjecture, 67–69
Completion, 8, 10
Complex number, 12, 26, 32, 52, 53, 57
Compulsion schema, 33, 35

Conceptual metaphor, 5, 8–11, 25–28, 31, 32, 41, 42, 50–53, 56, 57, 60–66, 73, 75, 78–80, 88, 90
Constancy, 35–37
Container image schema, 3, 11, 12, 28, 42, 44, 50, 63
Container schema, 26, 28, 29, 31, 32, 59, 61, 74, 78, 79
Continuity, 10, 26, 35–37
Coordinate system, 8, 23, 24, 33, 35, 52, 57
Counterforce schema, 33
Culture, 4, 6, 7, 15, 75, 76
Cycle schema, 26, 32, 35

D
Dehaene, S., 1, 3, 7, 14, 76, 89, 90
Devlin, K., 1, 15, 58, 75
Diagonal proof, 9, 29, 60
Diagrams, 4, 5, 11, 12, 23, 25, 31, 41, 42, 51, 52, 64, 73–90

E
Embodied cognition, 89
Euler, L., 38, 39, 41, 63, 81, 83
Existential graph, 4, 5

F
Figure-ground, 35, 37
Force schema, 33, 35, 60

Four-color theorem, 84, 85
Fractions, 11, 26, 29, 34, 35, 37, 75, 76
Framing, 49, 53–56

G
Galileo, 8, 9, 11, 23, 42
Gauss, C.F., 43
Gestalt psychology, 37
Gestures, 4, 7, 8, 27, 53, 76, 77
Gilmore, C., 4, 16
Gödel, K., 9–11, 17, 29, 60, 61, 82
Graphs, 4, 5, 12, 33, 40, 41, 63, 66, 68, 80, 81, 83–85
Graph theory, 35, 38, 39, 41, 42, 44, 45, 51, 63, 81, 83–86
Grounding metaphor, 11, 12

H
Hedblom, M., 3, 4, 25, 31, 32, 86, 87

I
I Ching, 44
Image schema, 2–6, 11–13, 16, 18, 23, 25–38, 42–45, 49–55, 57, 60–68, 73–83, 86–88, 90
Image schema theory, 1, 2, 5, 6, 17, 23–45, 49, 63, 73, 74, 77, 80, 85, 87–90
Imaginary numbers, 12, 27, 52
Incompleteness, 9, 10, 37
Infinity, 6, 8–10, 12, 23, 26, 29, 35, 39, 42, 43, 66

J
Johnson, M., 1, 5, 25, 32, 50, 59

K
Königsberg Bridges Problem, 66, 81, 83

L
Lakoff, G., 1–18, 25, 26, 28, 32, 49, 50, 52, 53, 59–63, 65, 73, 74, 90
Langacker, R.W., 32
Layering, 7, 8, 49, 53, 57–61, 64, 67, 88
Learning, 4, 7, 16, 25, 55, 56, 59, 64, 73–90
Leibniz, G.W., 44
Liar Paradox, 10, 69

Linearity, 6, 7, 12, 23, 66
Linguistics, 1, 2, 4, 5, 13, 25, 32, 44, 45, 49, 53, 65, 86, 89, 90
Linking metaphor, 10–13, 33
Looping, 49, 67–69

M
Manipulatives, 6, 74
Mapping, 6, 8–11, 13, 25, 26, 28, 31, 33, 34, 41–43, 49–53, 57–64, 67, 75, 77, 81, 88
Mathematical cognition, 1–5, 45
Mathematical education, 6
Mathematics and language, 2, 13–16, 89
Metaphor, 3, 5, 7–13, 16, 17, 25–29, 31–33, 41, 42, 50–53, 56, 57, 59–66, 73–75, 78–80, 84, 88–90
Metonymy, 26, 35
Modular arithmetic, 43
Montessori method, 28

N
Networks, 39–42, 50, 51, 61, 63, 65–67, 80–83, 87, 89, 90
Neural network, 41, 49, 85, 86, 89
Neuroscience, 1, 17, 73
Nonhuman number sense, 4
Number line, 3, 5–10, 12, 13, 23, 24, 28, 29, 32, 35, 37, 42, 43, 45, 51–57, 59–62, 64–66, 80
Number sense, 3, 4, 14, 15, 74–77
Numeration, 6, 74, 76
Núñez, R., 1–18, 25, 26, 28, 31, 32, 52, 60–63, 73, 75

O
Orientation schema, 26, 33

P
Path image schema, 3, 12, 23, 38, 51, 61, 64, 65, 67, 68, 80
Peirce, C., 4, 5, 62
Piaget, J., 4, 15, 44, 75
Plato, 17, 50
Poincaré, H., 63
Popper, K., 25
Pre-math, 4, 6, 14, 34, 56, 64, 74, 75, 90
Proximity, 35, 36

Psychology, 1, 5, 34, 35, 37, 41, 44, 80
Pythagoras, 29–31, 44, 51

Q
Quantity, 4, 6, 15, 28, 29, 33–35, 59, 74, 79

R
Race Course Paradox, 39
Radial network, 65–67

S
Schema, 1, 23, 49, 73
Set theory, 3, 12, 26, 29, 35, 42, 50, 60
Similarity, 29, 35–37, 89
Source domain, 5, 8, 10, 11, 25, 32, 38, 50–52, 57, 60–62, 64, 77, 88
Square number, 29, 30, 65
Stasis-motion-force, 38–41, 88
Subitizing, 4, 35, 74–76
System thinking, 41–44, 58

T
Target domain, 5, 8, 23–25, 38, 50, 51, 57, 58, 60, 63–65, 81, 88
Theory-making, 83
Topology, 13, 32, 34, 66
Triangular number, 30, 31, 65

V
Verticality schema, 5, 27, 52, 57
Vygotsky, L.S., 41, 75

W
Wallis, J., 3, 5, 6, 42, 45, 64
Where Mathematics Comes From, 1, 4, 10, 16, 25, 32, 73, 88

Z
Zeno of Elea, 38

Printed in the United States
by Baker & Taylor Publisher Services